Chirality effects in thermotropic and lyotropic nematic liquid crystals under confined geometries

Von der Fakultät Chemie der Universität Stuttgart zur Erlangung der Würde eines Doktors der Naturwissenschaften (Dr. rer. nat.) genehmigte Abhandlung

Vorgelegt von

M. Sc. Clarissa Franziska Dietrich

aus Wuppertal

Hauptberichter: Prof. Dr. Frank Gießelmann

Mitberichter: Prof. Dr. Jan Lagerwall

Tag der mündlichen Prüfung: 19. Juli 2019

Institut für Physikalische Chemie der Universität Stuttgart

2019

Bibliografische Information der Deutschen Nationalbibliothek

Die Deutsche Nationalbibliothek verzeichnet diese Publikation in der
Deutschen Nationalbibliografie; detaillierte bibliographische Daten sind im Internet
über http://dnb.d-nb.de abrufbar.

1. Aufl. - Göttingen: Cuvillier, 2019
   Zugl.: Stuttgart, Univ., Diss., 2019

   D 93

Cover:

Polarized Micrograph of the twisted polar director configuation formed by an achiral
lyotropic micellar nematic phase under capillary confinement

© CUVILLIER VERLAG, Göttingen 2019
   Nonnenstieg 8, 37075 Göttingen
   Telefon: 0551-54724-0
   Telefax: 0551-54724-21
   www.cuvillier.de

   ISBN 978-3-7369-7128-8
   eISBN 978-3-7369-6128-9

Declaration of Authorship

I hereby certify that the dissertation entitled "Chirality effects in thermotropic and lyotropic liquid crystals under confined geometries" is entirely my own work except where otherwise indicated. Passages and ideas from other sources have been clearly indicated.

Ich versichere, dass ich die vorliegende Arbeit mit dem Titel "Chirality effects in thermotropic and lyotropic liquid crystals under confined geometries" selbstständig verfasst und keine anderen als die angegebenen Quellen und Hilfsmittel benutzt habe; aus fremden Quellen entnommene Passagen und Gedanken sind als solche kenntlich gemacht.

Clarissa F. Dietrich

Publications

The experimental parts of this thesis were mostly carried out at the Institute of Physical Chemistry at the University of Stuttgart, Germany. The parts for which a cleanroom facility was needed were carried out in the MC2 cleanroom of the Chalmers University of Technology in Gothenburg. The results of this work were presented in several talks and posters at national and international conferences as well as in two scientific publications.

Publications:

1) P. Rudquist, C. F. Dietrich, A. G. Mark and F. Gießelmann. "Chirality Detection Using Nematic Liquid Crystal Droplets on Anisotropic Surfaces", *Langmuir* **2016**, *32*, 6140-6147.

2) C. F. Dietrich, P. Rudquist, K. Lorenz and F. Gießelmann. "Chiral Structures from Achiral Micellar Lyotropic Liquid Crystals under Capillary Confinement", *Langmuir* **2017**, *33*, 5852-5862.

Conference contributions:

1) C. F. Dietrich, A. Mark, F. Schörg, P. Rudquist, P. Fischer and F. Gießelmann. "Chiral induction of cholesteric phases by helical nanoparticles", *42nd German Liquid Crystal Conference* (O10), Stuttgart, Germany, **2015**.

2) C. F. Dietrich, P. Rudquist, K. Lorenz and F. Gießelmann. "Observation of chiral structures from achiral micellar lyotropic liquid crystals under capillary confinement", *2nd German British Liquid Crystal Conference* (O15), Würzburg, Germany, **2017**.

3) C. F. Dietrich, P. Rudquist, K. Lorenz and F. Gießelmann. "Observation of chiral structures from achiral micellar lyotropic liquid crystals under capillary confinement", *10th Liquid Matter Conference* (P3.030), Ljubljana, Slovenia, **2017**.

4) C. F. Dietrich, P. Rudquist and F. Gießelmann. "Confinement-enhanced chiral induction in lyotropic liquid crystals", *45th German Liquid Crystal Conference* (P17), Luxembourg, Luxembourg, **2018**.

5) C. F. Dietrich, P. Rudquist and F. Gießelmann. "Chiral structures from achiral micellar lyotropic liquid crystals under capillary confinement", *XXII Conference on Liquid Crystals Chemistry, Physics and Applications* (Invited I-10), Jastrzębia Góra, Poland, **2018**.

Acknowledgments

Many people helped and supported me during my doctorate and have made this dissertation possible and my time pursuing it wonderful. Remembering back to the beginning of my doctorate, I would first like to thank Prof. Dr. Frank Gießelmann, Prof. Dr. Sven Lagerwall and Prof. Dr. Jan Lagerwall for the organization of the Bandol Summer School 2014. One could not wish for a better start into research than learning all the basics of liquid crystals at the wonderful Côte d´Azur. Besides the excellent scientific education and lots of interesting discussions – sometimes with a marvelous view of the sunset – the Bandol Summer School gave me for the first time the opportunity to connect with scientists from all over the world, with some of them I have maintained a close friendship since then.

Furthermore, my special thanks go to:

- Prof. Dr. Frank Gießelmann for the opportunity to investigate a fascinating topic in liquid crystal research, his expert guidance and moreover his constant and invaluable extensive support
- Prof. Dr. Jan Lagerwall for preparing the second assessment for this thesis and for numerous helpful and inspiring discussions during conferences
- Prof. Dr. Sabine Laschat for taking over the post of chairperson in the examination
- Prof. Dr. Per Rudquist for starting to work with me on this topic (our collaboration goes back to the time when I did my master thesis and was working on helical nanoparticles), for proofreading my thesis and his extensive supervision over the last years. There are no words to describe how much I´ve learned from him during the past years. I am very grateful for all the discussions and skype sessions we had, the publications we wrote and that I had the opportunity to work with him several times at the Chalmers University of Technology. I will keep my stay at Gothenburg in good memory.

- Prof. Dr. Peter Collings for visiting Stuttgart and working with me on the light scattering analysis for nematic liquid crystals and proofreading this chapter of my thesis. I am very grateful to him for helping me out with his extensive expertise. I would not have understood the theory and analysis behind that method without him.
- Prof. Dr. Thomas Sottmann for his expert knowledge in light scattering
- Prof. Dr. Peer Fischer for being my GRAD*US* mentor
- Dr. Zoey Davidson for his expert knowledge in soft matter physics and the numerous scientific (and non-scientific) discussions
- Everyone who took part in the scientific discussion concerning the results of this thesis
- All members of the workshops for mechanics and electronics as well as the technical staff for their fast and uncomplicated support
- My bachelor students Markus Keller, Kristin Lorenz and Nadine Schnabel for their participation in research projects
- My former bachelor adviser Dr. Johanna Bruckner for proofreading and from whom I've learned to work with lyotropics and as a scientist in general, e.g. to pay attention to details
- My fellow student and friend Carsten Müller for helping me out with all kinds of technical and computer/software-related questions, for helping me photograph my textures and for giving me a ride to the university and back home for several years
- All present and former members of the workgroup for the excellent atmosphere and their willingness to help in every respect: Friederike Knecht, Marc Harjung, Michael Christian Schlick, Dr. Johanna Bruckner, Iris Wurzbach, Carsten Müller, Christian Häge, Sonja Dieterich, Christina Abele, Sebastian Marino, Andreas Bogner, Boris Tschertsche, Frank Jenz, Gabriele Bräuning, Inge Blankenship, Dr. Nadia Kapernaum, Dr. Stefan Jagniella, Elisa Ilg
- My friends, my family and everyone else who accompanied and supported me throughout my studies and doctorate
- My boyfriend Dr. Tobias Steinle for his support, for getting a physicist point of view and for helping me out with writing a Python Script
- My parents without whom none of this would have been possible

Table of contents

1 Introduction

This study investigates chirality effects in thermotropic and lyotropic nematic liquid crystals under confinement. Chirality is a phenomenon in nature that has been attracting attention in all disciplines of natural science for a very long time. The notion was introduced by Lord Kelvin saying "I call any geometrical figure, or group of points, chiral, and say that it has chirality, if its image in a plane mirror, ideally realized, cannot be brought to coincide with itself.".[1] Therefore, an object is called chiral if it cannot be superimposed to its mirror image in the absence of rotation-reflection axes. On the contrary, an object is achiral when it contains an axis of rotation-reflection implying that one can end up with the same structure if one rotates it about an axis and reflect it in a plane perpendicular to that axis.

The most intuitive example of a chiral object is the human hand. The left and the right hands are mirror images of each other, which cannot be superimposed. Chiral objects are also referred to as being handed. Another very important example of chirality in nature can be found in biochemistry where most of the biomolecules and all essential amino acids are chiral. Chiral molecules, which differ only with respect to their handedness, are called enantiomers and are labeled D (Dexter, right) or L (laevus, left). In nature, only L-amino acids occur and can be metabolized by living beings on earth. The origin of this homochirality in nature is still an unsolved question and attracts researchers across all disciplines of natural science. The occurrence of chiral structures in a system containing only achiral components is called spontaneous mirror symmetry breaking and can add to this discussion and is by itself of fundamental interest.

In this study, we discovered new examples of mirror symmetry broken structures in the field of liquid crystals, which are obtained by means of the delicate interplay of topology, elastic free energy and interfacial anchoring conditions of liquid crystals in confined geometries. These systems allowed us to study chirality effects in a very sensitive way and to detect qualitatively and quantitatively tiny amounts of chiral additives in a range in which, e.g., only one out of 3000 molecules is chiral.

In this thesis, we present two new methods for chirality detection and sensing for two classes of liquid crystals: one that can be used for thermotropic liquid crystals, which are of broad commercial interest (LC displays), and one for lyotropic liquid crystals which are more life-science related and biologically compatible.

In order to introduce some fundamental concepts of liquid crystals, this chapter will first deal with the liquid crystalline state of matter in general and then focus on the characteristics of the simplest liquid crystalline phase, the nematic phase, and its chiral variant, i.e., the chiral nematic – so-called cholesteric – phase.

1.1 The liquid crystalline state of matter

The classical states of matter are usually summarized as solid, liquid, gas and plasma. Many other states are known to exist, such as glass or liquid crystal. In the 20th century, lots of additional states of matter were identified, such as superfluid and Bose-Einstein condensate, but none of these are observed under normal conditions. During the discovery of the liquid crystalline state, chirality played by the way an essential role because in 1888 Friedrich Reinitzer, who was a biologist, investigated the chiral substance cholesteryl benzoate, which appeared to him to have two melting points.[2] One year later, the physicist Otto Lehmann studied the appearance of two melting points by means of a polarizing optical microscope and coined these materials "liquid crystals". Therefore Reinitzer and Lehmann were both pioneers of the research field dealing with liquid crystals.[3]

Liquid crystalline phases, also so-called mesophases, combine properties of solid matter, like for example optical anisotropy, with qualities of a liquid, like for example fluidity.[4] The building blocks of a liquid crystalline phase are named mesogens. In a crystal, the three-dimensional lattice implies long-range positional order. The atoms or molecules are kept on their lattice sites by interactions resulting in additional long-range orientational order. Therefore, a crystal exhibits both long-range positional and orientational order.

Nevertheless, both long-range orders can occur separately from each other. The origin of anisotropic physical properties like birefringence lies in the long-range orientational order in the sense that these properties are directional and not the same in all three different spatial directions. In liquids, however, no long-range order exists, only a short-range order with respect to the neighboring particles.

The physical properties of a liquid are thus isotropic. Liquid crystals, as the name implies, combine the typical crystalline feature of the long-range orientational order giving rise to anisotropic physical properties with the fluidity of an ordinary liquid.

In Figure 1.1 a typical phase sequence of liquid crystalline material is shown schematically. Coming from the highly ordered crystalline lattice, a liquid crystalline material loses its long-range positional order in one, two and three dimensions, but is able to preserve the long-range orientational order upon heating towards the melting point. When increasing the temperature further to the clearing point, the long-range order vanishes completely and only the typical short-range order of a simple fluid persists which on the other hand itself disappears at the transition to the gaseous state of matter at the boiling point.

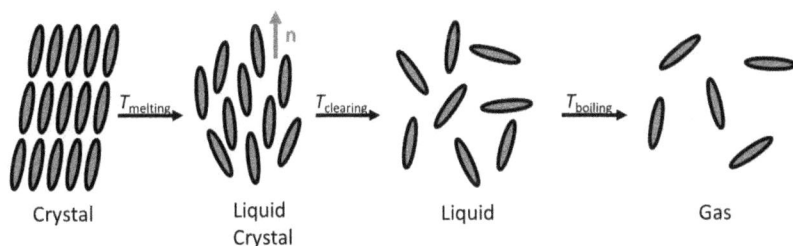

Figure 1.1: Exemplary phase sequence of liquid crystalline material. The mesogens are schematically drawn as rods. With increasing temperature the material passes through several phase transitions – marked by the melting, clearing and the boiling point – from highly ordered (low symmetry) to disorderd (high symmetry) phases. In the liquid crystalline phase, the director **n** indicates the long-range orientational order of the long molecular axes.

The basic requirement for a substance to exhibit a liquid crystalline phase is to consist of building blocks with a non-spherical shape. There are in principle two simple ways to implement that – either in a rod-like or in a disc-like fashion. Having, e.g., rod-like particles, the system can maximize translational entropy by aligning the long axes of the individual particles along with the spatial and temporal mean values of preferred orientation. This preferred orientation is called the director **n**. Due to the difference in the degree of order, various liquid crystalline phases can be formed. Some liquid crystalline phases may even exhibit an additional one- (smectic) or two-dimensional (columnar) long-range positional order of the mesogenic centers.

In general two different kinds of liquid crystals can be distinguished. The first class is the thermotropic liquid crystals which are well known and which have been first recognized as a new state of matter. Their liquid crystalline phase appearance is solely caused by temperature variations (at constant pressure). The building blocks, in that case, are simple organic molecules of anisotropic shape. Rod-shaped mesogens typically have flexible alkyl- or alkoxy chains attached to an elongated aromatic core. The ordering of liquid crystalline phases occurs when molecular and steric interactions (e.g. aromatic core-core interactions with flexible alkyl chains hindering crystallization), leading to parallel alignment of neighboring molecules, dominate over orientational entropy.

The second class is lyotropic liquid crystals, the appearance of which can be dated back to an age in which the soap-making process was developed.[5] Lyotropic liquid crystals are at least two-component systems consisting of surfactant molecules or polymers and a solvent, in most cases water. The surfactant molecules, which are amphiphilic organic molecules having a hydrophobic alkyl chain and a hydrophilic head group, assemble themselves if surrounded by, e.g., water into aggregates which – if of non-spherical shape – are the building blocks of a lyotropic phase. These aggregates are called micelles. The solvent concentration is the crucial parameter for the formation of a lyotropic liquid crystalline phase. In comparison to thermotropic liquid crystals, temperature plays only a minor role.[6]

From an application point of view, one should first mention the use of thermotropic liquid crystals in modern display technology (LCDs = liquid crystal displays). However, lyotropic liquid crystals are in almost the same manner important for applications, like, e.g., in the detergent and cosmetic industry and for medical and pharmaceutical use.[6] It is of basic interest to know the solvent concentration at which lyotropic liquid crystalline phases occur because they influence product properties like viscosity, stability or dispersing and foaming power. For cosmetics, it is good to know how to use surfactants as surface-active and emulsifying agents and stabilizers. With respect to medical and pharmaceutical use, lyotropic mesogens are part of in-vitro hydrophilic drug delivery systems (e.g. Nicotinamide) as they can pass through lamellar double layers which leads to the biological significance of lyotropic liquid crystals, given the fact that the eukaryotic cell membrane exhibits a lyotropic-lamellar layer structure. Additionally, it was found that the DNA in the cell nucleus assembles in a lyotropic-hexagonal structure and that DNA also forms a cholesteric phase at certain concentrations in water. Hence, the life-scientific relevance of lyotropic liquid crystals should not be underestimated.[7,8]

Furthermore, chirality plays an important role in liquid crystal applications, e.g. the RealD 3D system used for stereoscopic film projection takes advantage of chiral liquid crystal configurations simultaneously transmitting left-eye images with circularly polarized light of one handedness and right-eye images with the light of the opposite handedness. This technology allows viewers to til their heads without compromising image separation.[9] In addition, optical thermal sensors are based on cholesteric LCs.[10–12] Last, but not least, achiral liquid crystals, which exhibit spontaneous chiral symmetry breaking, can be used as sensors for molecular chirality. Being achiral they will show configurations of either handedness with equal probability, but only a very small amount of chiral disturbance will tip over this delicate balance favoring configurations of one handedness over the other. An example could be that such a system uses the ratio of domains with left- and right-handed twists in disclination lines as a measure of chirality.[13]

1.2 The nematic phase

1.2.1 Order parameter and anisotropic properties

According to the spatial and temporal mean values, the mesogens locally align themselves along a preferred direction in the nematic phase (N). This way, the system minimizes its free volume, together with its free energy by gaining translational entropy at the expense of reducing orientational entropy. The preferred direction is called the director **n**. In Figure 1.2 a simulated snapshot of this simplest liquid crystalline phase with rod-shaped particles is shown and illustrates that the director **n** just reflects the mean orientation of the long axes of all particles; a long-range positional order as it is the case in a crystalline lattice does not exist.

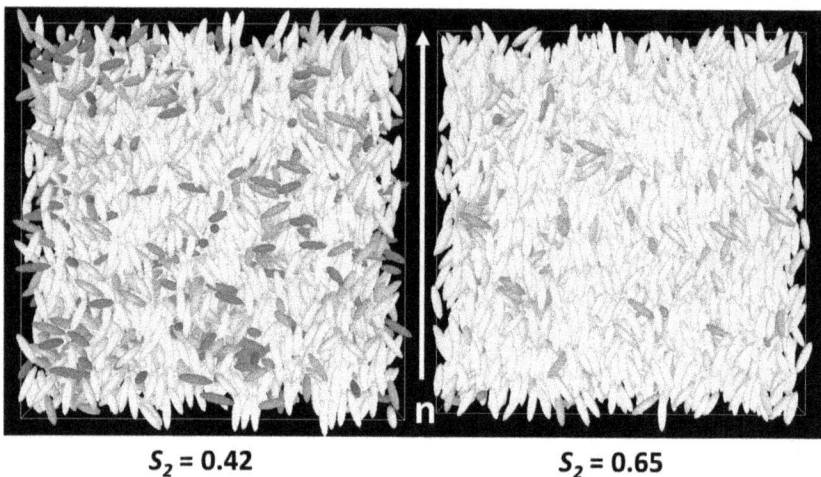

$$S_2 = 0.42 \qquad\qquad S_2 = 0.65$$

Figure 1.2: Snapshot of a nematic phase with calamitic mesogens for two different values of the orientational order parameter simulated by Christian Häge. Yellow rods indicate that the angle α between the molecular long axis and the director **n** is small, whereas orange and red rods indicate a larger angle α.

A measure for the quality of orientational order of the mesogenic main axes along the director **n** is quantified by the orientational order parameter S_2.[14] It takes into account how much, on average, the principal axis with the highest symmetry of every single mesogen differs from the director **n** described by the angle α_i between those.

The orientational order parameter has to meet several requirements, e.g., that it is 0 in the isotropic phase where the particles are randomly oriented and that it equals 1 in a perfectly orientationally ordered system, e.g., all rods are exactly parallel to each other. The orientational order parameter can be described as:

$$S_2 = \frac{1}{2} \langle 3\cos^2\alpha_i - 1 \rangle \ . \tag{1}$$

For a typical nematic phase, S_2 is in the range between 0.4 and 0.7 and values also depend upon temperature. It is to be mentioned that the directions $+\mathbf{n}$ and $-\mathbf{n}$ are physically equivalent, meaning that there is no macroscopic polarity although the mesogens themselves can be polar. S_2 can be denoted also as $\langle P_2(\cos\alpha_i) \rangle$ following from the series expansion of the orientational distribution function in terms of Legendre polynomials.[14]

For a uniaxial phase with the principal symmetry axis along z, all tensorial properties χ along x and y are the same such that $\chi_{11} = \chi_{22} \neq \chi_{33}$. The corresponding 2nd rank tensor is written as:[15]

$$\chi_{\alpha\beta} = \begin{pmatrix} \chi_\perp & 0 & 0 \\ 0 & \chi_\perp & 0 \\ 0 & 0 & \chi_\parallel \end{pmatrix} \ . \tag{2}$$

If one takes the magnetic susceptibility as an example, $\chi_\perp$ and $\chi_\parallel$ are the susceptibilities perpendicular and parallel to the director, respectively. As the tensor consists of only two components, it can be divided into two parts, the mean value $\langle\chi\rangle = 1/3\ (\chi_\parallel + 2\chi_\perp)$ and the anisotropic part $\Delta\chi = \chi_a = \chi_\parallel - \chi_\perp$.

In the case of disc-shaped nematics (N_D) the magnetic susceptibility is negative, meaning that the alignment of the director perpendicular to an external magnetic field is favored in terms of free energy. On the other hand, calamitic shaped nematics (N_C) have a positive diamagnetic anisotropy trying to align the director parallel to an applied magnetic field. Concerning birefringence, an N_C phase has, in general, a positive birefringence whereas an N_D phase exhibits negative birefringence. The optical anisotropy of liquid crystals will be explained in more detail in Chapter 3.2 in combination with the concept of polarized optical microscopy.

The anisotropic part of the tensor introduced in Equation (2) can be expressed as:[15]

$$\chi^a_{\alpha\beta}=\chi_{\alpha\beta}-\langle\chi\rangle\delta_{\alpha\beta}=\begin{pmatrix}\chi_\perp & 0 & 0\\ 0 & \chi_\perp & 0\\ 0 & 0 & \chi_\parallel\end{pmatrix}-\begin{pmatrix}\langle\chi\rangle & 0 & 0\\ 0 & \langle\chi\rangle & 0\\ 0 & 0 & \langle\chi\rangle\end{pmatrix}$$

$$=\begin{pmatrix}-\dfrac{1}{3}\chi_a & 0 & 0\\ 0 & -\dfrac{1}{3}\chi_a & 0\\ 0 & 0 & \dfrac{2}{3}\chi_a\end{pmatrix}.\qquad(3)$$

In order to get rid of the physical dimensions, the anisotropy χ_a is normalized by the maximal anisotropy, which is possible in the case of an ideal alignment in a crystalline solid at absolute zero temperature, rendering the order parameter tensor $Q_{\alpha\beta}$:[15]

$$Q_{\alpha\beta}=\frac{\chi^a_{\alpha\beta}}{\chi^{max}_{\alpha\beta}}=\frac{\chi_a}{\chi^{max}_a}\begin{pmatrix}-1/3 & 0 & 0\\ 0 & -1/3 & 0\\ 0 & 0 & 2/3\end{pmatrix}=S_2\left(n_\alpha n_\beta-\frac{1}{3}\delta_{\alpha\beta}\right).\qquad(4)$$

The ratio χ_a/χ^{max}_a is equivalent to S_2 from Equation (1) and represents the scalar amplitude of the order parameter indicating the degree of molecular statistical order. On the other hand, the tensor $Q_{\alpha\beta}$ gives the orientational part of the order parameter. Because the director **n** is nonpolar, the expression of the quadratic combination $n_\alpha n_\beta$ is used.

According to Landau and Lifshitz, the free energy is a function of temperature and of the order parameter.[16] If the order parameter is sufficiently small, the free energy can be expanded in a power series of the invariants of Q up to the fourth rank. Taking into account the free energy of the isotropic phase F_{iso} the Landau free energy F of a nematic liquid crystal can be written as:[17]

$$F=F_{iso}+\frac{1}{V}\int d^3r\left[\frac{a}{2}Q_{\alpha\beta}Q_{\beta\alpha}-\frac{b}{3}Q_{\alpha\beta}Q_{\beta\gamma}Q_{\gamma\alpha}+\frac{c}{4}\left(Q_{\alpha\beta}Q_{\beta\alpha}\right)^2\right].\qquad(5)$$

The coefficients b and c can be regarded as temperature-independent, whereas close below the isotropic-nematic phase transition temperature T_C (clearing temperature) the coefficient a has to change sign at the lower absolute stability limit (supercooling limit) of the isotropic phase at $T=T^*$ ($<T_C$). This is taken into account by the relation $a=a_0(T-T^*)/T^*$.

1.2.2 Elastic free energy

Equation (5) is valid in the case that the orientational tensor $Q_{\alpha\beta}$ does not vary in space. However, this assumption does not hold if confined geometries come into the game and defects and disclinations occur. Possible distortions by external forces, e.g., boundary forces, mechanical stress, electric or magnetic fields, have to be taken into account by a gradient elastic energy $F_{elastic} = F_{elastic} (Q(r), \nabla Q(r))$. The increase of the free energy is described by the continuum theory which originally was devised by Oseen[18] and Zocher[19] and further developed by Frank[20] into its nowadays well known mathematical form. The Frank-Oseen free energy which can describe distortions in the director field is:

$$F_{elastic} = \int d^3 x \left[\frac{1}{2} K_1 (\nabla \cdot \mathbf{n})^2 + \frac{1}{2} K_2 (\mathbf{n} \cdot (\nabla \times \mathbf{n}))^2 + \frac{1}{2} K_3 (\mathbf{n} \times (\nabla \times \mathbf{n}))^2 \right]. \tag{6}$$

Here, $\mathbf{n}$ denotes the (local) director, K_1 is the splay elastic constant, K_2 is the twist elastic constant and K_3 is the bend elastic constant. The nematic elastic moduli are always positive and have the dimension of a force, in the range of 10^{-12} Newton.

In general, the three elastic constants have the same order of magnitude and therefore a "one constant approximation" is often applied. In addition to these three major elastic moduli, there is the saddle-splay elastic constant K_{24}, which becomes important only for particular situations, in which a distortion has a two- or three-dimensional structure such as nematic droplets in an isotropic fluid or the blue phases.[15]

In Figure 1.3 the main three elastic director distortions, which can occur in a bulk nematic liquid crystal, are drawn schematically in case of disk- and rod-shaped building blocks. It is pointed out that solely the twist deformation is chiral in the sense that the twisted director field lacks mirror symmetry. Note that in two dimensions, a splay deformation of a N_C looks like a bend deformation in N_D and vice versa.

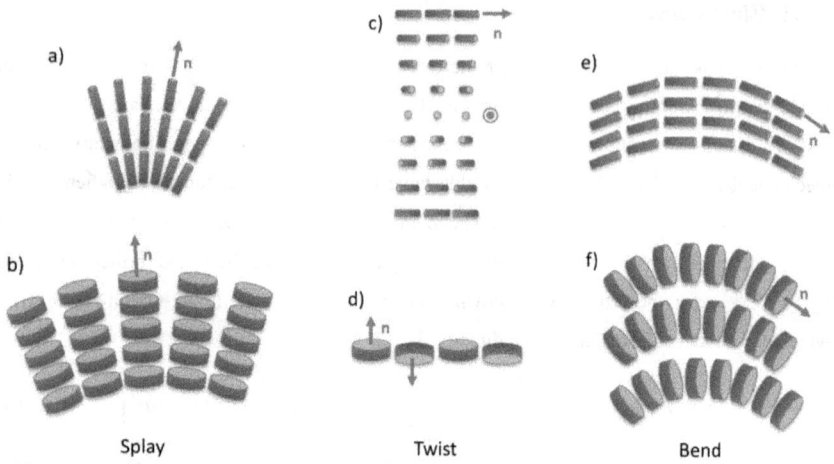

Figure 1.3: The blue rods and discs depict the rod- or disc-shaped building blocks of a N_C or a N_D phase, respectively. The director **n** is shown as a red arrow. From left to right: Splay, twist and bend distortion configurations. (a) – (b) and (e) – (f) show splay and bend deformations which are non-chiral, but twist deformations in (c) – (d) are chiral; in the case shown here, it is right-handed. Note that in two dimensions, a splay deformation of a N_C, see (a), looks like a bend deformation in N_D, see (f), and vice versa.

1.2.3 Defects and disclinations

The concept of defects stems from crystallography in which defects are disruptions of the ideal crystalline lattice such as so-called vacancies (point defects) and dislocations (line defects) appearing due to broken translational symmetry. Topological defects in nematic liquid crystals appear during the symmetry breaking of the isotropic to nematic phase transition and originate particularly in the breaking of the rotational symmetry. Due to the fact, that all orientations of the director **n** are equally probable, it is possible that in different places different orientations are spontaneously formed and these domains merge while growing. At the interface of these domains distortions and discontinuity of the director field within the volume can form defects and disclinations.

Another possibility of how defects can be generated consists of dirt particles. If a nematic liquid crystal is placed between two glass plates which align the director parallel to the surface area and this surface area is not perfectly clean, dirt particles disturb the homogeneous alignment of the director being the seed for defect formation. The areal distribution of defects results in a so-called Schlieren texture which is characteristic for the nematic phase when looking through a polarized optical microscope. This texture consists of point and line defects connected by dark brushes building up a network.[21] The best, but definitely not the easiest way, to define and classify defects is by homotopy groups saying that in a uniaxial phase the appearance of point and line defects are allowed. A more descriptive way of explanation is that defects represent singularities of the director field which is a spot where the order parameter breaks down and the director orientation is undefined.[15,22,23]

In Figure 1.4 a typical polarized optical micrograph of a characteristic nematic Schlieren texture is shown. Schlieren textures occur if nematics are placed between two untreated glass substrates and a perfect homogenous parallel alignment of the director on the glass surface cannot be achieved; instead, the orientation varies slowly in the plane of the substrate. Another nematic texture in which defects play an important role is the so-called thread-like texture which originally gave the nematic phase its name ("nema" is the Greek word for "thread"). The dark lines in this texture are disclination lines which either connect two two-fold defects or form closed loops.[21]

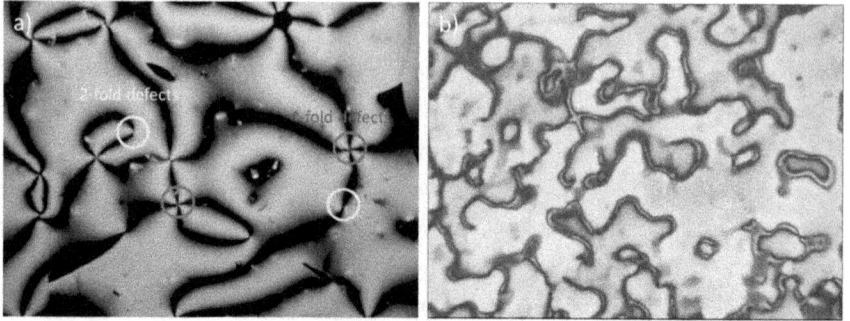

Figure 1.4: (a) Polarized optical micrograph of a typical Schlieren texture of a nematic phase with 4-fold and 2-fold singularities.[24] (b) Thread-like texture of the nematic phase.[21]

The singularities of the director field can be classified by their strength s and their dimension D. Ranking defects according to their dimension, 0-dimensional (point), 1-dimensional (line) and 2-dimensional (wall) defects can be distinguished. The strength of a defect reflects the degree of rotational continuity. The director is traced along a closed loop around the defect and the number of rotations which the tip has to perform doing a full circle determines the strength s. The sign of the strength is determined by the fact whether the rotation was performed clockwise or counter-clockwise. Simply speaking, the number of brushes coming from a defect has to be divided by four and the defect is of positive charge when the brushes rotate along with the rotation of the polarizers and of negative charge when rotating the other way round. Given that, 2- and 4-fold defects result in possible defect strengths s of $\pm \frac{1}{2}$ and ± 1. Figure 1.5 shows the surrounding director fields of these defects and their corresponding appearance under crossed polarizers. It is to be mentioned that all these disclinations are so-called wedge disclinations because the singularity is perpendicular to the surrounding director field. However, another type is the so-called twist disclinations which have the disclination line parallel to the local director field.[22] A wedge disclination is transformed into a twist disclination through a director rotation about a certain axis normal to the disclination line, see Figure 3.17a,b.

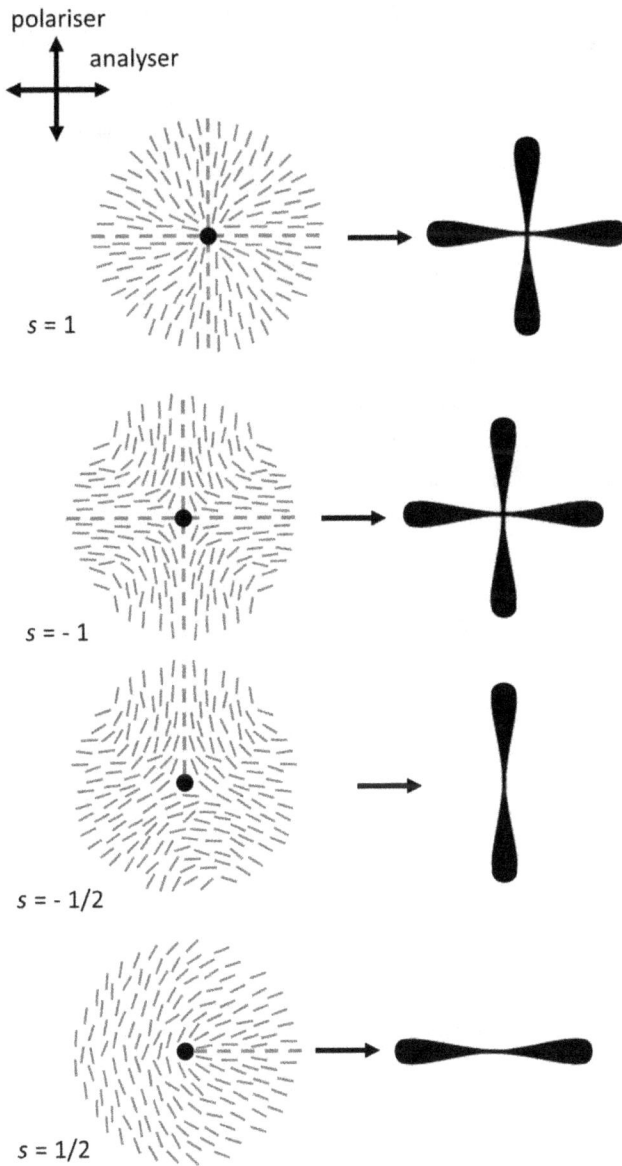

polariser

analyser

$s = 1$

$s = -1$

$s = -1/2$

$s = 1/2$

Figure 1.5: Schematic drawing of the director fields in the vicinity of ± ½ and ± 1 defects and how the brushes look like under crossed polarizers in the polarized optical microscope. When rotating the polarizers the punctual singularity remains on its place whereas the brushes rotate continuously.[24]

Defects of the same strength, but opposite sign, can annihilate with each other like electric charges of opposite sign do, generating a defect-free director field. Defects which are not of the same strength cannot annihilate with each other, but they can form another singularity which sums up the strengths of the original defects. The elastic energy stored around a disclination per its unit length, the so-called line tension, is proportional to s^2 which means that it is energetically favored to split up an $s = \pm 1$ defect into a pair of $s = \pm \frac{1}{2}$ defects:[15]

$$F_{\text{disclination}} = \pi K s^2 \ln\frac{r_{\max}}{a} \quad . \tag{7}$$

The limits for the integration of the free energy are given by $r_{\max}$ which is the sample radius and a is the core of the disclination which is excluded from consideration, see Figure 1.6. K is the elastic modulus in the one constant approximation. The energy of a disclination per unit length diverges logarithmically when $r \rightarrow \infty$. Nevertheless, this condition is not realistic because there are additional confinements due to other defects for example. Typically $r_{\max} \approx 10 - 100 \, \mu m$, $a \approx 10 \, nm$, $\ln(r_{\max}/a) \approx 10$, $F_{discl} \approx 30 \, K \approx 3^{-10} \, Jm^{-1}$.[15]

Figure 1.6: Schematic illustration of the parameters $r_{\max}$ which is the sample radius and a being the core of the disclination which is excluded from consideration.

The interaction energy W_{12} per unit thickness of two disclinations of s_1 and s_2, separated by the distance r_{12}, can be expressed by:[15]

$$W_{12} = -2\pi K s_1 s_2 \ln\frac{r_{12}}{a}. \tag{8}$$

The force of interaction is proportional to $1/r_{12}$. This reflects the analogy to the force of interaction of two parallel wires carrying electric currents. It also demonstrates that disclinations of opposite sign attract each other because W_{12} is positive and decreases with shrinking r_{12}.

In products of optical technology, defects reduce the performance, e.g., defect walls in twisted nematic cells. However, in order to identify different phases and to help understand complex three-dimensional periodic structures like the blue phase of cholesterics, defects and disclinations are a very practical tool.[20] Furthermore, elastic properties can be analyzed by investigating defect interactions.[20]

1.3 The cholesteric or chiral nematic phase

The cholesteric phase N* is the chiral version of the nematic phase introduced in the previous chapter and is traditionally called cholesteric because, at first, it has been observed in cholesteryl esters. Chirality is the lack of mirror symmetry and derives from the Greek word for "hand". Similar to the nematic phase the N* phase exhibits solely orientational order of the long molecular axis, but in contrast to the N phase a macroscopic helical superstructure having a twist axis perpendicular to the local director is observed. The periodicity of the helical superstructure with a full rotation of 360° of the director is called the pitch P and its inversed value P^{-1} is called twist. If $P < 0$ the helical superstructure is left-handed; if $P > 0$ it is right-handed. The nematic phase has point group $D_{\infty h}$. Adding chirality reduces this symmetry to D_{∞}. According to de Vries, a chiral nematic phase with infinite pitch has the same structure as the achiral nematic phase.[25] However, even though the pitch of a chiral nematic is infinite, it is still a chiral system whereas the achiral nematic phase remains a non-chiral system. In Figure 1.7 the director field of a chiral nematic LC is shown. The local director is indicated, the pitch P is denoted and the twist axis is shown in red. The Frank elastic free energy introduced in Equation (6) has to be modified for the chiral nematic phase according to: [15,20]

$$F_{\text{elastic}} = \int d^3 x \left[\frac{1}{2} K_1 (\nabla \cdot \mathbf{n})^2 + \frac{1}{2} K_2 (\mathbf{n} \cdot (\nabla \times \mathbf{n}) + q_0)^2 + \frac{1}{2} K_3 (\mathbf{n} \times (\nabla \times \mathbf{n}))^2 \right]. \qquad (9)$$

With the helical wave vector $q_0 = 2\pi/P$ and $q_0 > 0$ for a right-handed and $q_0 < 0$ for a left-handed twist. The pitch ranges typically about $0.1 - 10$ µm and the inverse of the pitch, the twist, is a measure for chirality.[4]

There are different possibilities for how to obtain a chiral nematic phase. Either the mesogens are chiral themselves or a chiral dopant which induces chirality is added to the non-chiral nematic host. In the case of lyotropic liquid crystals, the use of chiral surfactants or chiral solvents can also give a cholesteric phase. When having a chiral dopant added to the system, the sign of the induced pitch depends on the enantiomer which was used. For example, if (R)-

mandelic acid would induce a left-handed helical superstructure in the nematic host, the (*S*) enantiomer would give a right-handed helix. The handedness depends on the used chiral dopant as well as on the used host phase.

In Figure 1.7 the helical superstructure of a chiral nematic phase consisting of disc-shaped building blocks and the corresponding typical Fingerprint texture which can be observed under crossed polarizers are illustrated. The Fingerprint texture is characterized by a periodic bright-dark stripe pattern that originates from the alternating orientation of the director rotating between homeotropic and planar orientation. Homeotropic orientation means parallel alignment of the director to the path of light giving a dark picture of the birefringent sample under crossed polarizers. In the case of planar alignment, the director is oriented perpendicular to the path of light, giving a bright picture in a polarized optical microscope. The periodicity of the stripe pattern represents a half-pitch length by a director rotation about 180°.

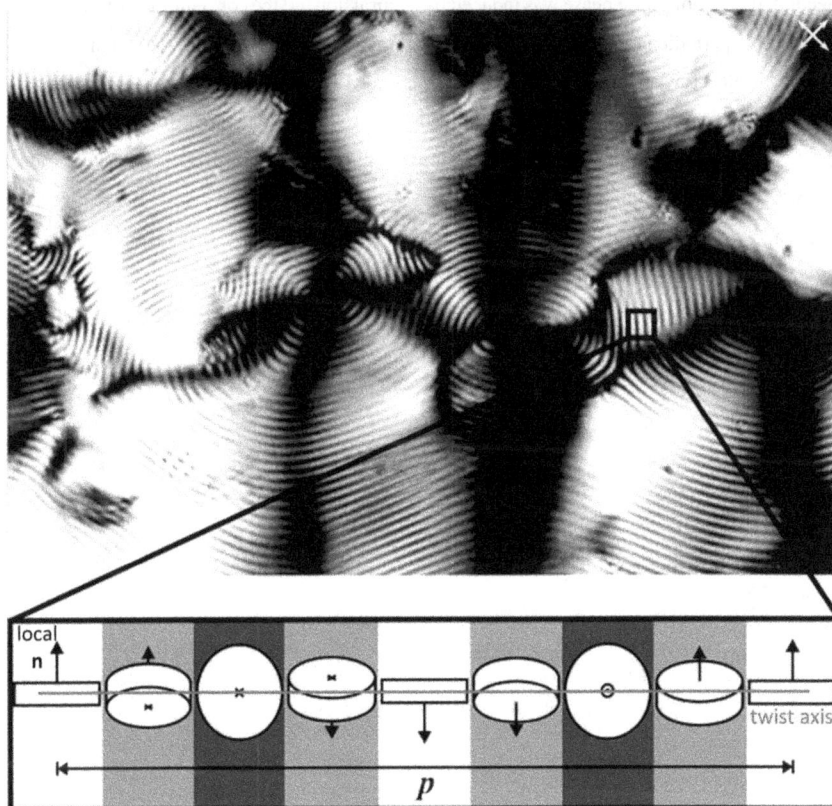

Figure 1.7: Helical superstructure of a chiral N_D phase and the resulting Fingerprint texture observed under a polarized optical microscope. The arrows indicate the orientation of the local director and the pitch P is denoted. The twist axis is shown in red. Viewing direction is from the top. Picture reprinted with permission from ref. [26].

The pitch of a cholesteric phase depends on different parameters. First, there is the temperature dependence. If the N* phase transforms at lower temperature directly into a crystalline phase, a linear change of the pitch with temperature is often found.[27–29] However, if the low-temperature phase is layered (smectic), a hyperbolic behavior towards this phase transition temperature can be observed. In literature, this behavior is explained as a pre-transitional effect.[30–34] Which kind of temperature dependency holds is according to Osipov connected to the different mechanisms of chiral induction.[35]

Second, if the cholesteric phase was induced by a chiral dopant, the concentration of this dopant affects the periodicity of the induced helical superstructure. Typically the pitch diverges hyperbolically towards zero dopant concentration (Figure 1.8a) because at zero dopant concentration the N* phase becomes a normal N phase with infinite pitch. If the inverse pitch, the twist P^{-1} is plotted against the molar fraction x of the chiral dopant the dependency becomes linearly proportional going through the origin (Figure 1.8b). In Figure 1.8 the diagrams of the pitch and the inverse pitch versus the dopant concentration are shown exemplarily for a lyotropic nematic system doped with (R)-mandelic acid.

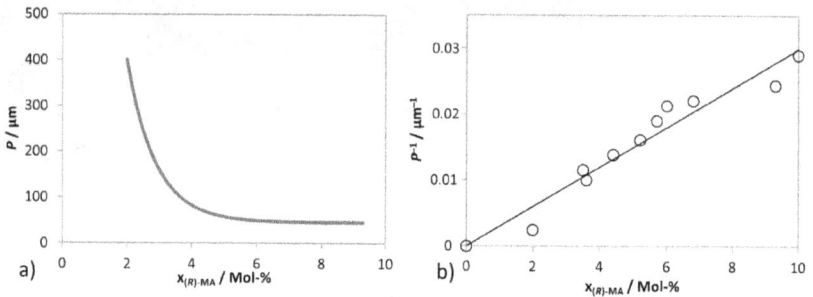

Figure 1.8: (a) Pitch P and (b) Twist P^{-1} plotted against the molar fraction of (R)-mandelic acid $x_{(R)\text{-MA}}$ in the lyotropic nematic system CDEAB/DOH/H$_2$O for a mass ratio of CDEAB/DOH = 6.6. Diagrams are redrawn based on ref. [36].

However, this linear dependence is only valid for small dopant concentrations. When increasing x_{dopant} further, the ascent of the twist flattens out and saturates at a certain value.[37] This behavior is illustrated schematically in Figure 1.9. In the linear regime at low dopant concentrations, the slope of the linear regression represents the *Helical Twisting Power HTP*.[l] The *HTP* is a measure for the ability of a chiral dopant to induce a chiral nematic phase in the achiral host material.

It is a characteristic and specific value for the chiral dopant and the host phase and is defined as:[38–40]

$$H = \lim_{x \to 0} \left(\frac{\partial P^{-1}}{\partial x} \right)_T .$$ (10)

---

[l] To avoid misapprehension, the *Helical Twisting Power* is abbreviated as *HTP* in the continous text and and as *H* in equations.

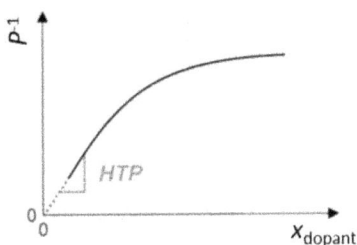

Figure 1.9: Schematic illustration of the correlation between the twist P^{-1} and the molar fraction of the dopant x_{dopant}. For low dopant concentrations the twist increases linearly with increasing x_{dopant}, the corresponding slope represents the *Helical Twisting Power HTP*. At higher dopant concentrations the curve saturates.[41]

Having a high *HTP*, the chiral dopant effectively induces a small pitch at even low dopant concentrations into the nematic host phase. For enantiomers of different handedness, the *HTP* has the same absolute value, but opposite sign. This means that the *Helical Twisting Power* of a dopant is a chiral indicator.[42,43] A negative *HTP* indicates that the chiral dopant induces a left-handed helix, whereas a positive *HTP* indicates that the chiral dopant induces a right-handed helix in the host phase.

1.4 Lyotropic micellar and chromonic liquid crystals

In the introduction of the liquid crystalline state of matter, it has already been mentioned that in general two kinds of liquid crystals exist. On the one hand, there are the thermotropic liquid crystals where the organic molecule itself represents the anisotropically shaped mesogen (rod-like, disc-like, banana-like, etc.). And on the other hand, there are the lyotropic liquid crystals where the building blocks correspond to non-spherical supramolecular assemblies dispersed in a solvent, typically water. This means that lyotropic liquid crystals are at least a two-component system. The super-molecular assemblies can be formed by different means like standard amphiphilic surfactant molecules forming micelles (lyotropic micellar liquid crystals), or polymers dissolved in a solvent (lyotropic liquid crystal polymers) occurring in nature in the DNA, the spider silk or polysaccharides, or non-spherical nanoparticles dispersed in a solvent, or disc-shaped drugs or dye molecules stacking on each other to a cylinder in hydrophilic surroundings (lyotropic chromonic liquid crystals).

Going back to the origin of formation of a liquid crystalline phase in general, it can be said that according to Onsager there are two factors that come into play.[44] There are on the one hand the intermolecular and steric interactions (aromatic core-core interactions leading to anisotropic dispersion interactions and flexible alkyl chains hindering crystallization), as well as entropic effects like sacrificing orientational freedom to gain translational entropy when aligning e.g. rod-shaped building blocks parallel to each other.

In the Onsager model of identical rigid rods, an isotropic to nematic phase transition can occur if the volume fraction Φ of the rods in the system is larger than a certain threshold value where L and D are the length and the diameter of the rods, respectively:[44]

$$\Phi > \Phi_{nematic} = \frac{3.3D}{L} \ . \tag{11}$$

This relation is temperature independent, meaning that solely the volume fraction $\emptyset$ can be the tuning parameter for the formation of a liquid crystalline orientational order. This leads us to the discussion which of these two factors dominates in the formation of a thermotropic or lyotropic liquid crystalline phase.

The Maier-Saupe Theory is the standard theory for thermotropic nematic LCs which is based on anisotropic dispersion interactions and the polarizability of the mesogens. In this theory, temperature is the key parameter. On the contrary, lyotropic LCs are commonly described by the Onsager Theory, in which the steric interactions between rigid rods lead to parallel orientation of the mesogens. This minimizes the excluded volume and the translational entropy is raised at the expense of the orientational entropy. This effect is only dependent on the concentration of the mesogens and their anisotropy (ratio between length L and diameter D), but not on the temperature.

Let us start with the lyotropic micellar liquid crystals, which are the most typical and common representatives. As the name already indicates, the building blocks of these systems are anisotropically shaped micelles. In order to form micelles, the system must contain surfactant molecules and a solvent. Sometimes also a cosurfactant can be added in order to change the shape of the micelles. A cosurfactant cannot form micelles on its own.[6] A typical surfactant molecule and a schematic illustration of how an N_D phase is formed are shown in Figure 1.10.

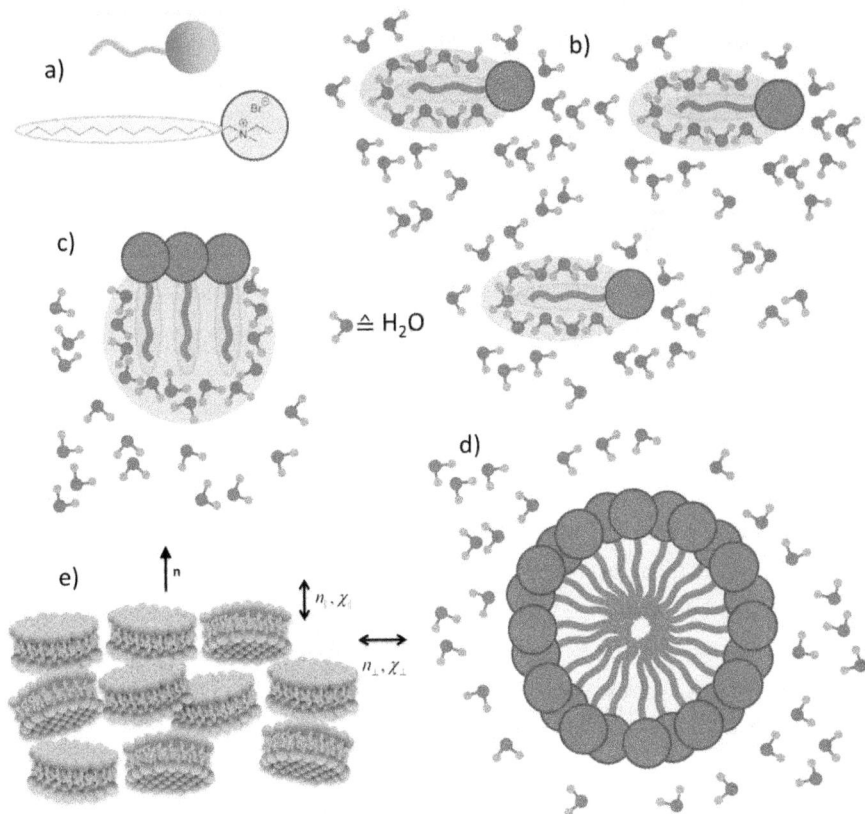

Figure 1.10: (a) Molecular structure of the surfactant N, N-dimethyl-N-ethylhexadecyl-ammonium bromide (CDEAB) and corresponding schematic drawing of the amphiphilic surfactant with the polar head group (blue) and the alkyl chain (gray) are shown. (b) A solution of amphiphiles in water. The amphiphile molecules force the surrounding water molecules to form highly ordered cages around each alkyl chain which brings a high loss in entropy. (c) An aggregation of hydrophobic tails reduces the number of water molecules in ordered cages which reduces the entropy loss. (c) Due to the formation of micelles, all hydrophobic tails are separated from the surrounding water and the ordered shell of water molecules is minimized. This reduces the entropy loss further. (d) Schematic illustration of a lyotropic N_D phase; refractive indices parallel $n_{\parallel}$ and perpendicular $n_{\perp}$ as well as magnetic susceptibilities parallel $\chi_{\parallel}$ and perpendicular $\chi_{\perp}$ to the director **n** are indicated. Note that the edges of the micelles are covered with polar head groups as well. Figure (a) and (e) are reprinted with permission from Langmuir from ref. [45]. Figure (b) – (d) are redrawn based on ref. [46].

In order to form micelles, the surfactant molecules have to exhibit an amphiphilic structure with a hydrophobic (alkyl) chain and a hydrophilic head group. The polar head group can contain different functional groups which are, e.g., anionic, cationic, amphoteric or nonionic. On the other hand, the surrounding solvent plays a crucial role in the micellar formation process.[6]

The further discussion exemplifies water as the solvent because it is most often used. A schematic overview of the micellar formation process is shown in Figure 1.10. Water has different features that come into play, e.g., the high polarity and the capability to form strong intermolecular hydrogen bonds ($\Delta H \approx 20$ kJ/mol in pure water). In the crystalline phase, the tetrahedral lattice has a relatively low density and this structure is partly maintained in the liquid phase. Therefore, there is a lot of free volume between this tetrahedral arrangement giving the water molecules numerous possibilities of allowed positions. This means the entropy of this system is very large.

By introducing amphiphilic molecules, the water structure can rearrange to a certain amount in a way that encapsulates the nonpolar parts of the molecules from the surrounding polar water, see Figure 1.10b. Simultaneously the polar head groups of the amphiphilic molecules form hydrogen bonds with the surrounding water; the binding energy can range from 10 – 50 kJ/mol.[6] However, for the surrounding water molecules, this water caging goes with a loss of entropy, because their configuration has to get more ordered.[47] This is called "hydrophobic effect" and is the dominant reason why amphiphilic molecules can only be dissolved as monomers at low concentrations. An increase of the solubility can only be achieved if the loss in entropy of the water molecules can be compensated. First, this is realized by an adsorption film of the surfactant molecules going to the air-liquid interface. The hydrophilic head groups point into the water and the hydrophobic chains stick out into the air, see Figure 1.10c. This film decreases the surface tension of water. Additionally, another way to increase the solubility by lowering the contact area of the hydrophobic alkyl chain with water is the formation of aggregates where the hydrophobic parts are encapsulated in the interior space and the hydrophilic head groups constitute the surficial area of the so-called micelle, see Figure 1.10d. The aggregation of amphiphilic molecules into micelles is determined by a critical surfactant concentration, the *critical micelle concentration CMC*.[5,6] Above this concentration, it is not possible for the system to adsorb any more molecules at the air-liquid interface; therefore, the micelle formation process begins. Micelles are no statically determined structures, the formation process is reversible and a balance between aggregated and monomer solubilized amphiphilic molecules is set.

The micellar shape depends on the concentration of the surfactant and the molecular structure thereof. This dependence is quantified in the so-called packing parameter which relates the effective volume of the amphiphile to the length of the hydrophobic chain and the cross-section area of the polar head group. Especially the size of the hydration sphere around the polar head group is in the end strongly affecting the micellar shape. The packing parameter can be changed by concentration and the addition of cosurfactants.[6] The aggregation number gives the number of particles a micelle consists of. With constant aggregation numbers and increasing surfactant concentration, the number of micelles increases. Due to that and inter-micellar interactions, structures with long-range order can build up, ranging from solely long-ranged orientational order in the nematic phase (N_C and N_D) to additional 1D long-ranged positional order in the lamellar phase L_α and even 2D long-ranged positional order in the hexagonal phase H_α. A schematic overview of the correlation between surfactant concentration and the corresponding lyotropic liquid crystalline phases is shown in Figure 1.11.

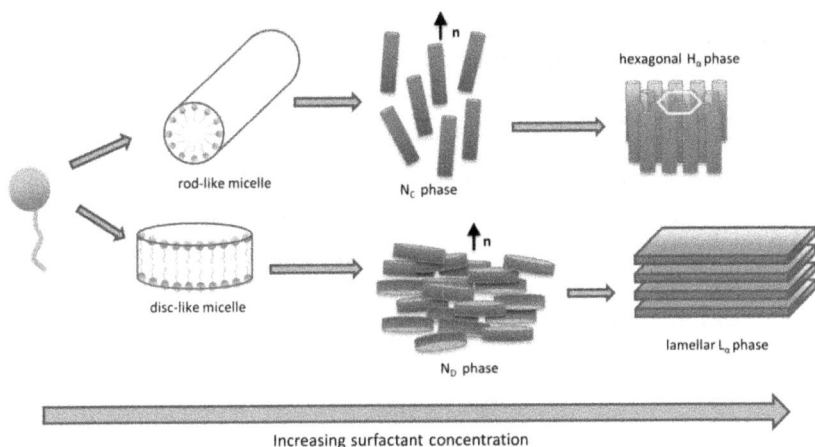

Figure 1.11: Schematic overview of the correlation between surfactant concentration and the lyotropic liquid crystalline phases. The sketch is redrawn based on ref. [6].

In chromonic liquid crystals disc-shaped molecules with a polyaromatic center and ionizable groups at the outer part form cylindrical stacks due to $\pi - \pi$ interactions. These cylindrical super-molecular assemblies represent the building blocks of a lyotropic chromonic liquid crystal.[48-50] Dissolved in water, the counter ions of the ionizable groups (often Na^+) are free dispersed in the surrounding water, whereas the cylindrical stacks are being left negatively charged.

They align themselves in a rod-like nematic fashion like it is shown in Figure 1.12 or even in a smectic layer structure. Typical examples of disk-shaped molecules that can form chromonic liquid crystals are the antiasthma drug disodium cromoglycate (DSCG), the food and textile dyes Sunset Yellow (SSY), Allura red, Methyl orange.

The formation of the cylinders depends upon concentration on the one hand and temperature on the other hand. At lower temperatures, the length of aggregates increases leading to the formation of a N_C phase with long-range orientational order.[51] A way to express that is the energy, which is needed to break one aggregate into two:[52]

$$E = E_0 - E_r \approx 10 \, k_B \, T \tag{12}$$

with E_0 being the attraction energy between the aromatic cores and E_r being the electrostatic repulsion energy between the ionized groups at the outer part.

With constant temperature and dye concentration but augmented ionic concentration, the repulsion between molecules within the aggregates and between the aggregates themselves is decreased and longer aggregates form and a more ordered phase is supported. On the contrary, adding, for example, NaOH leads to an increased disassociation of sodium ions, which increases E_r and therefore destabilizes the ordered phase. Another property, which can be tuned by the ionic concentration, is the flexibility of the cylinders. For example, a bend deformation brings negative surface charges closer together and a higher ionic concentration could screen possible electrostatic repulsions making the aggregates more flexible.[51,53] Similar features of chromonic liquid crystalline behavior is found in double-strain DNA assemblies and the ion concentration becomes important for example in the case of DNA wrapping around nucleosomes.[8,54,55]

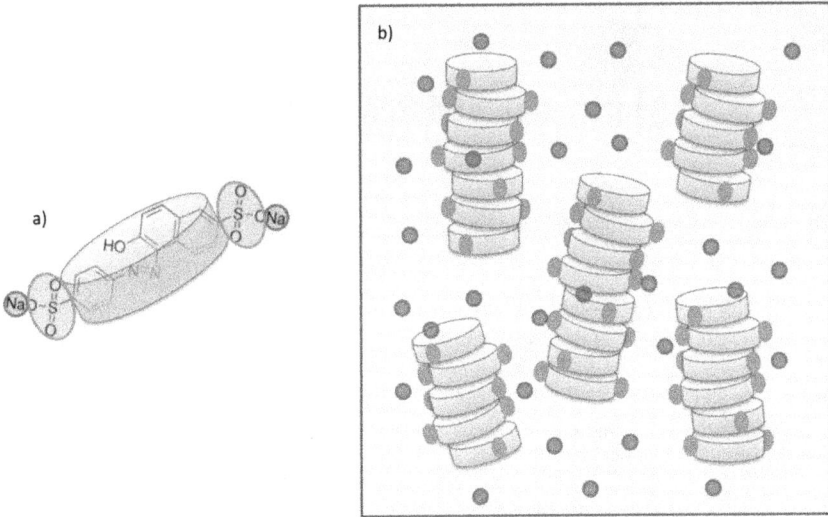

Figure 1.12: (a) Molecular structure of Sunset Yellow, the flat aromatic center is marked as blue disc, the ionizable sulfate groups are marked in green and the sodium counter ions are marked as red dots. (b) Schematic illustration of cylindrical assemblies of SSY dissolved in water, forming an N_C phase. The sodium ions are disassociated, leaving the aggregates with a negative charge. The picture is redrawn based on ref. [56].

2 Motivation and Scope of this thesis

The numerous applications of liquid crystals originate from the high responsivity of liquid crystals to various stimuli, such as temperature, optical radiation, magnetic and electric fields, the surrounding confinement as well as chirality. The combination of the latter two and their interaction is the subject of this thesis.

Chirality is a phenomenon in nature that attracts attention in all disciplines of natural science for a very long time, as the origin of homochirality in biochemistry. Adding a chiral dopant to a nematic phase, the microscopic chirality of the chiral molecules is transferred to the macroscopic scale by inducing chirality in the liquid crystalline host phase in terms of a helical superstructure with a certain handedness and a characteristic periodicity.[57,58] This periodicity, the so-called pitch, is typically in the range of micrometers.[37] In order to understand the still unclear process of chiral induction further, the study of weakly doped systems having pitch lengths in the range of several milli- or centimeters is inevitable. In general, director configurations of confined liquid crystals reveal basic physical properties due to a delicate interplay of topology, elastic free energy, and interfacial anchoring conditions.[59] This interplay can be influenced very sensitively by the addition of chirality.

The goal of this thesis is to investigate chirality effects in thermotropic and lyotropic nematic liquid crystals by exploiting the sensitive director fields within confined geometries providing a potential amplification of the effects of interest. Suitable confining geometries have to be found in order to observe qualitatively and to measure quantitatively chirality effects for very small amounts of chiral additives in LCs. Recently, in the special case of achiral nematic chromonic lyotropic liquid crystals chiral director configurations were reported for various confining geometries, representing examples of mirror symmetry breaking.[60-66] The formation of these unusual chiral configurations was attributed to a surprisingly small twist elastic modulus which was found to be one order of magnitude smaller than the splay and bend moduli.[56,67,68]

These findings lend themselves to be investigated whether similar chiral configurations in the case of achiral standard micellar lyotropic LCs can occur due to a potentially similar anomaly in the elastic constants, which is to be verified. Furthermore, this leads to the issue whether those configurations can serve as extremely sensitive tool to study the process of chiral induction like, e.g., the transition to homochirality – because in the achiral system the left- and right-handed twist sense occur with the same probability, this energetic degeneracy is lifted by adding a chiral dopant.

Thermotropic liquid crystals do not exhibit such elastic peculiarities like the one found in the special case of chromonic lyotropic LCs. Therefore, none of these chiral director configurations can be observed in thermotropic LCs. Thus, some different geometrical confinement for investigating chirality effects of weakly doped thermotropic liquid crystals has to be found. In this thesis, two new methods to study chiral induction for both classes of liquid crystals – for thermotropic as well as for lyotropic LCs – are investigated. In addition, the impact of the geometrical confinement is discussed, e.g., how the confinement amplifies, induces, and influences the detection of chirality.

3 Chiral structures of achiral micellar lyotropic liquid crystals under capillary confinement

Chapter Overview

The spontaneous formation of chiral structures in a system consisting of exclusively achiral components is known as spontaneous mirror symmetry breaking and of fundamental interest across all disciplines of natural science. Recently, in the field of liquid crystals, the appearance of spontaneous reflection symmetry broken configurations in achiral chromonic liquid crystals under capillary confinement was reported.[61] These observations were attributed to a small twist elastic modulus which is one order of magnitude smaller than the splay and bend moduli. In this chapter, the observation of similar chiral configurations in the case of a classical, achiral micellar lyotropic liquid crystal is demonstrated. Similarities and differences to the case of chromonic liquid crystals are discussed, in particular, the conditions under which spontaneous mirror symmetry breaking occurs. This part of my thesis is mainly based on my publication with Per Rudquist, Kristin Lorenz and Frank Gießelmann on "Chiral Structures of Achiral Micellar Lyotropic Liquid Crystals under Capillary Confinement" which appeared 2017 in Langmuir.[45] Furthermore, the addition of a chiral dopant to this highly chiral sensitive system is investigated.

3.1 Director field configurations under capillary confinement

This introduction focuses on nematic director configurations under capillary confinement with homeotropic boundary conditions in general. Before introducing the chiral configurations which were recently found in chromonic liquid crystals, let´s first have a look at the well-known achiral director configurations of nematic liquid crystals within cylindrical geometries.[69–72] An overview of these non-chiral configurations is shown in Figure 3.1.

Having the nematic director aligned perpendicularly to the inner glass surface of the capillary the most intuitive director field would adopt in a radial fashion which only requires splay deformation. This so-called planar radial (PR) configuration leads to a high frustration of the director field along the axis of the capillary yielding a disclination line of the topological strength $s = +1$. However, the +1 disclination is energetically not stable towards splitting up into two disclination lines of strength $s = + \frac{1}{2}$. This is due to the fact, that the elastic energy stored around a disclination per unit length – the so-called line tension – is proportional to s^2, see Equation (7) in Chapter 1.[15] This results in the planar polar (PP) configuration which is illustrated in Figure 3.1b. The director field of the PP configuration is characterized by two + $\frac{1}{2}$ disclination lines which are located close to the capillary walls providing the opportunity to retain an essentially undistorted homogeneous director field in the central area of the capillary. The elastic distortions of the director field in the PP configuration involve splay and bend deformations.

Another possibility of how the +1 disclination line in the PR configuration can be avoided is in a continuous way meaning an escape of the director field into the third dimension.[69,73,74] This is what mostly happens when a nematic liquid crystal is confined to a capillary and called escaped radial (ER) configuration. By the continuous transformation of the director field involving splay and bend deformations along the disclination line, the extended defect core is removed leaving two point defects at the two outer ends of the capillary such that the overall topology strength is preserved. Overall, the majority of the director field is defect-free and the ER configuration is thus more stable than the PR given that the radius of the capillary is much larger than the molecular dimensions. The escape directions can be either to the left or to the right or even change within the capillary through additional point defects. Having lots of point defects separating the different escape directions, the configuration is then called escaped radial with point defects (ERPD). The typical ER configuration is shown schematically in Figure 3.1c.

Which of those three configurations occur in a specific case depends on the capillary radius and the explicit values of the elastic constants and their ratios, e.g., with decreasing capillary radius the PP configuration is energetically favored with respect to the ER configuration.[75]

All these three configurations are non-chiral because they have mirror symmetry. Jeong et al. recently discovered that by filling lyotropic chromonic liquid crystals into cylindrical capillaries chiral configurations in which this reflection symmetry is broken can be obtained.[61] There is, on the one hand, the chiral analog to the ER configuration, the so-called twisted escaped radial (TER) configuration, see Figure 3.2. And on the other hand, there has been a twisted version of the planar polar configuration observed which was labeled as twisted planar polar (TPP), see Figure 3.4. Additionally, if having planar anchoring at the inner glass surface of the capillary, a third chiral configuration was found, the escaped twist (ET) configuration.[64,66] However, the ET configuration is not explained in more detail because the following study focuses on the configurations under homeotropic anchoring to the inner glass surface.

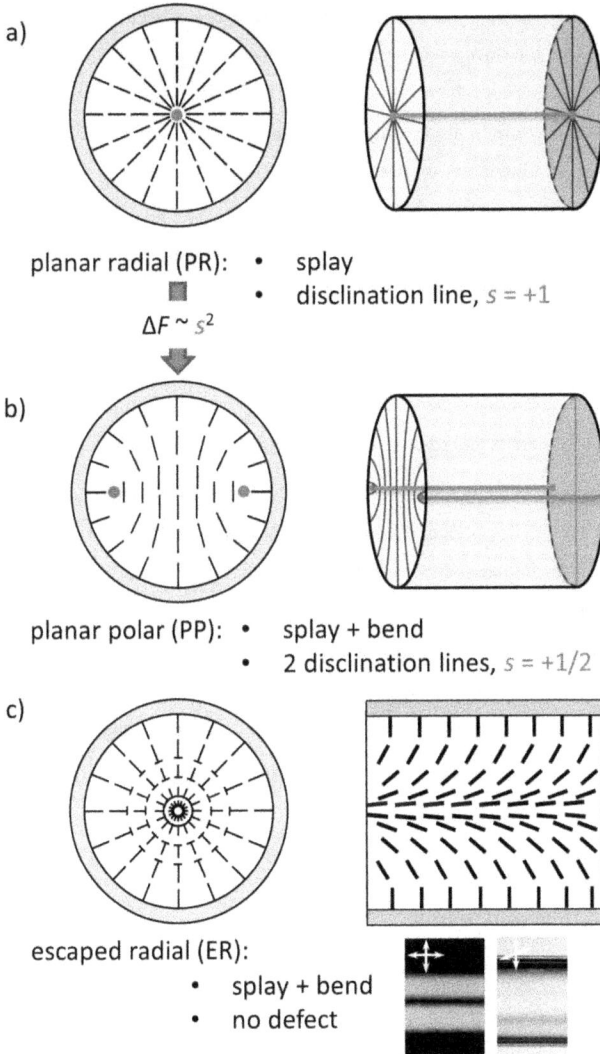

a)

planar radial (PR): • splay
■ • disclination line, $s = +1$
$\Delta F \sim s^2$

b)

planar polar (PP): • splay + bend
• 2 disclination lines, $s = +1/2$

c)

escaped radial (ER):
• splay + bend
• no defect

Figure 3.1: Overview of non-chiral nematic director field configurations under capillary confinement with homeotropic boundary conditions: (a) Planar radial (PR), (b) planar polar (PP) and (c) escaped radial (ER) configurations. The disclination lines are indicated in red. The PR and PP configurations have one $s = +1$ disclination line and two $s = +\frac{1}{2}$ disclination lines, respectively, which are marked in red. Only the ER configuration is continuous without a defect in the bulk, which can be verified by decrossing the polarizers, see inset of polarized optical micrographs. Figure from ref. [45], reprinted with permission from Langmuir.

Figure 3.2a shows the director field of the TER configuration from the side view of a capillary. Figure 3.2c,d shows the director field in the case of weak and strong homeotropic anchoring respectively. In the TER configuration, some twist is added to the pure splay-bend director field of the non-chiral ER configuration, comparing Figure 3.1c and Figure 3.2a. The superimposed twist goes along all directions of the capillary diameter. Under crossed polarizers this is indicated by a brightening of the dark brush along the axis of the capillary, compare inset of Figure 3.1c and Figure 3.2b. Due to the fact that it is still a non-chiral material that exhibits these mirror symmetry broken configurations, the energies of the two twist handedness are degenerated and domains of opposite twist sense can occur with equal probability. Chiral macroscopic domains with opposite twist senses can be observed within one capillary. A more detailed characterization of the here exemplarily shown polarized optical micrograph (Figure 3.2b) is given in the results and discussion part of this chapter.

This TER configuration was so far only found in lyotropic chromonic liquid crystals (namely the system of sunset yellow SSY and water) and studied in detail by Jeong et al. who attributed this phenomenon to the peculiarities of chromonics having a remarkably small twist elastic constant, which is one order of magnitude smaller compared to the splay and bend constants.[61] Due to that, strong splay and/or bend deformation can energetically escape into twisting which leads to equilibrium helical structures. In a similar way, an unusually small saddle-splay modulus gives stabilized chiral configurations upon planar anchoring, like in the case of the above mentioned escaped twist configuration.[64,66]

Figure 3.3 is reprinted with permission from PNAS from ref. [61] and shows the energy diagram of the elastic free energy of the TER configuration. The diagram demonstrates that the director field of the TER can minimize its total elastic free energy by adding twist elastic free energy. ΔF is the energy difference between TER and ER configuration for the individual contributions from splay, twist and bend elastic energies as well as their sum. As calculated under the assumption that the splay constant equals the bend constant, i.e. $K_1 = K_3 = K$, the formation of a TER configuration becomes energetically favored when the twist elastic constant decreases below a critical value of $K_2 < K_2^c \approx 0.27K$.[61]

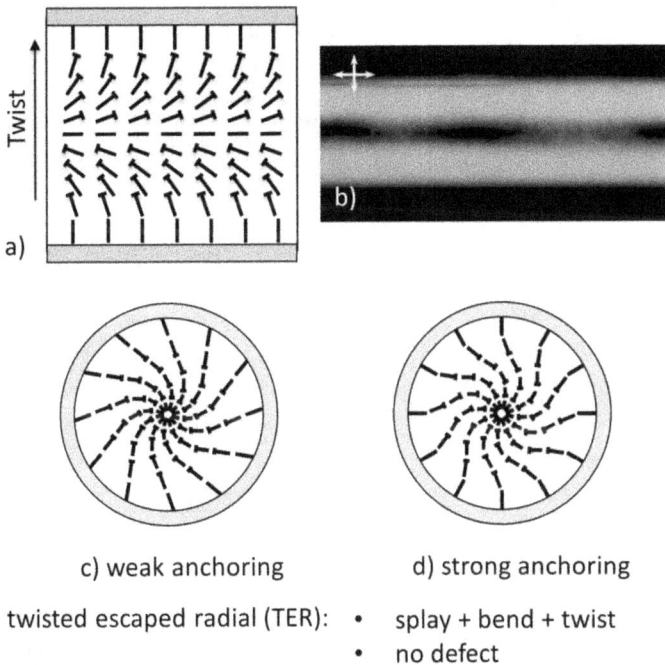

c) weak anchoring   d) strong anchoring

twisted escaped radial (TER):   • splay + bend + twist

• no defect

Figure 3.2: Director field of the twisted escaped radial (TER) configuration: (a) Side view, the twist axis of the TER configuration goes perpendicular to the cylinder axis. (b) TER regions appear bright and orange under crossed polarizers. (c) Cross-section of the capillary in the case of weak homeotropic anchoring. (d) Cross-section in the case of strong homeotropic anchoring. Figure from ref. [45], reprinted with permission from Langmuir.

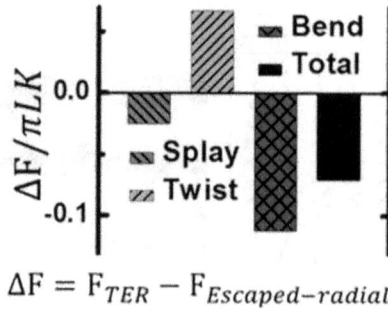

$$\Delta F = F_{TER} - F_{Escaped-radial}$$

Figure 3.3: Free energy diagram of the TER configuration showing that the director field of the TER minimizes its total elastic free energy by adding twist elastic free energy. ΔF represents the difference in elastic free energy of the individual contributions as well as their sum from splay, twist and bend elastic energies in the TER configuration (F_{TER}) with respect to the ones in the non-twisted escaped radial configuration ($F_{Escaped-radial}$). The numerical calculations were based on the assumption $K_1 = K_3 = K$ and L is the length of the cylinder. The diagram is reprinted with permission from PNAS from ref. [61].

Figure 3.4 shows the director field of the TPP configuration, as suggested by Jeong et al.[61]. It was observed – as well as the TER configuration – in the nematic lyotropic chromonic liquid crystal SSY under capillary confinement.[61] The TPP is distinct from the non-twisted planar polar configuration by a twisting of the two $s = +\frac{1}{2}$ disclination lines along the axis of the capillary forming a double helix. This configuration – as well as the TER configuration – lacks mirror symmetry and is therefore chiral. Like in the TER configuration, the energy of left- and right-handed twist domains is degenerated. Overall, these two domains thus occur with the same probability. The reason why the chiral analog of the PP configuration is formed is suspected to lie – as discussed in the case of the TER – in the small twist elastic modulus of nematic chromonic SSY.[61]

Figure 3.5 is reprinted with permission from PNAS from ref. [61] and shows two polarized optical microscope images of the TPP configuration of the chromonic liquid crystal SSY confined to a cylindrical capillary under polychromatic illumination. In Figure 3.5a the TPP (left side) is replacing the TER configuration (right side). This reflects the formation process of the TPP: In the case of chromonic SSY, the double helices of the TPP nucleate at arbitrary positions in the capillary, often at the ends of the capillary, and start to grow. Two domains of opposite twist sense are separated by domain-wall-like defects, as shown in Figure 3.5b.

According to Crawford et al.[76], the energetics of the TER to TPP transition could be related to the saddle-splay modulus K_{24} and a certain anchoring strength of the alignment layer (the inner glass surface was polymer-coated to achieve homeotropic anchoring of the chromonic N_C phase). This hypothesis suggests that with weak anchoring strength a deviation from the homeotropic radial director orientation towards the inner capillary wall is allowed and surface disclinations can be formed. In the case of chromonic SSY, the two $s = +\frac{1}{2}$ disclination lines of the TPP are indeed located quite near the capillary walls.[61]

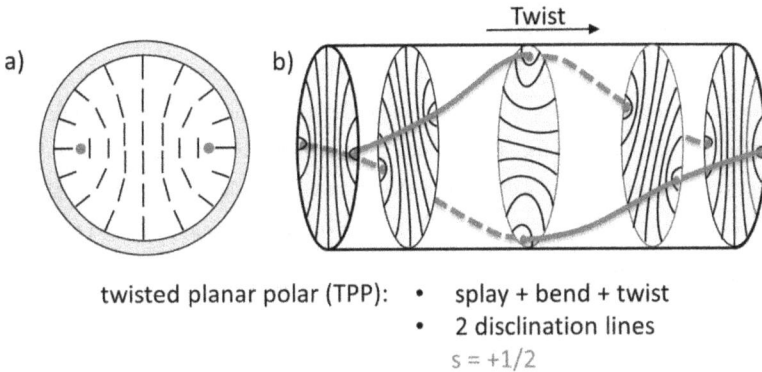

twisted planar polar (TPP): • splay + bend + twist
• 2 disclination lines
$s = +1/2$

Figure 3.4: Director field of the twisted planar polar (TPP) configuration as it was suggested in ref. [61]. (a) Cross-section of the capillary. (b) Side view, the two $s = +\frac{1}{2}$ disclination lines of the TPP configuration twist along the cylinder axis. Figure from ref. [45], reprinted with permission from Langmuir.

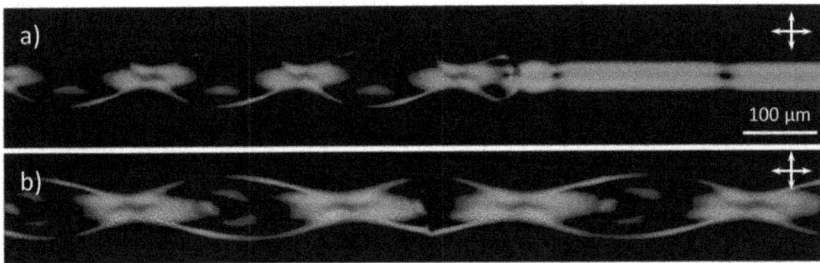

Figure 3.5: Polarized optical micrographs of the TPP configuration of the lyotropic chromonic liquid crystal SSY confined to a cylindrical capillary under polychromatic illumination and crossed polarizers. (a) Shows the formation process of the TPP (left side): the TPP nucleates at an arbitrary position and replaces the TER (right side), the two dark spots in the TER configuration are domain walls. (b) At the center of the image, a domain-wall-like defect separates double helices of opposite handedness. The pictures are reprinted with permission from PNAS from ref. [61].

3.2 Optical anisotropy of liquid crystals and polarizing optical microscopy

Optical anisotropy is a phenomenon that occurs when the velocity of light in a medium depends on the polarization plane of the electromagnetic wave in the material. The medium interacts differently with light regarding the orientation of the medium towards the polarization and propagation direction of the incoming light beam. A simple liquid has isotropic physical properties; therefore, the index of refraction is independent of the direction of the incident light beam. On the other hand, crystalline materials exhibit anisotropic physical properties, among which is the optical anisotropy meaning that their index of refraction depends on the light's direction of propagation as well as its polarization. The dependence on polarization is causing birefringence so that two perpendicular polarizations of the incident light propagate with different velocities and refract at different angles in the crystal. This gives rise to a split-up of the beam of light into an ordinary beam and an extraordinary beam introducing a phase shift. After leaving the medium, these two beams interfere with each other and become elliptically polarized light.

Optical anisotropy is a characteristic feature of liquid crystals.[77] The anisotropic shape of the mesogens evokes anisotropic physical behavior like it was introduced in Chapter 1.2.1. Liquid Crystals can be divided according to the number of optical axes; uniaxial liquid crystals only have one optical axis whereas a biaxial liquid crystal possesses two optical axes. For light propagating along an optical axis, the velocity does not depend on the polarization and no birefringence occurs, although the medium is overall anisotropic. This means that the optical axis indicates an axis of full rotational symmetry of the cross-section of the optical indicatrix (Figure 3.6b,f) meaning that the refractive index remains constant when rotating the medium around this axis.

The example given in Figure 3.6a shows the thermotropic liquid crystal n-(4-methoxybenzylidene)-4-butylaniline "MBBA". It has a typical rod-shaped molecular structure in which the electronic polarizability is different in the direction parallel and perpendicular to the long molecular axis. Transferring this from the microscopic to the macroscopic scale due to the long-range orientational order of the nematic phase, this gives rise to two different indices of refraction, which are denoted as $n_\parallel$ and $n_\perp$. However, these indices refer now rather to the director **n** than to the long axis of an individual molecule. In general, in the case of thermotropic liquid crystals, an N_C phase has a positive birefringence $\Delta n = n_\parallel - n_\perp > 0$, whereas an N_D phase exhibits negative birefringence $\Delta n = n_\parallel - n_\perp < 0$, see Figure 3.6a,e.[77]

In contrast, in micellar lyotropic liquid crystals an N_C phase is commonly optically negative and an N_D phase optically positive.[6] This is because thermotropic liquid crystals consist of organic molecules with an aromatic core along which the electronic polarizability is high whereas along with the non-polar hydrocarbon bonds it is small. For a rod-shaped organic molecule with an aromatic core that would mean that the refractive index $n_{\parallel}$ would be greater than the refractive index $n_{\perp}$ (Figure 3.6a). For a disc-shaped molecule with an aromatic core that would mean that the refractive index $n_{\perp}$ would be greater than the refractive index $n_{\parallel}$ (Figure 3.6e).

However, typical lyotropic liquid crystals consist of rod- or disc-shaped micelles formed by amphiphilic surfactant molecules with no aromatic core so they have smaller electronic polarizability to start with.[6] In case of a rod-shaped micelle, the refractive index $n_{\perp}$ is higher than the refractive index $n_{\parallel}$ because the electronic polarizability is always higher along the long axes of the molecules and in the case of a rod-shaped micelle, the long axes of the surfactant molecules go along the diameter of the micelle (Figure 3.6d). Additionally, for ionic surfactants, bulky and electron-rich counter ions could increase the polarizability. In case of a disc-shaped micelle, the long axes of the surfactant molecules go perpendicular to the diameter of the micelle, which means that the refractive index $n_{\parallel}$ is greater than the refractive index $n_{\perp}$ (Figure 3.6g). This results in an opposite optical anisotropy for an N_C and an N_D phase depending upon whether it is a thermotropic or lyotropic liquid crystal. It is worth pointing out that the optical anisotropy of the N_D and N_C phase in a lyotropic liquid crystal is the other way around if the surfactant molecules contain aromatic groups or perfluorinated alkyl chains such that the N_C phase becomes optically positive and the N_D phase optically negative, but this is more the exception than the normal case.[6]

When a light beam propagates through the optically anisotropic medium at an angle ζ with respect to the optical axis the incoming beam is, due to two different refractive indices, split into two components, an ordinary beam n_o and an extraordinary beam n_e, see Figure 3.6c. The refractive indices n_o and n_e are derived from the principal refractive indices $n_{\parallel}$ and $n_{\perp}$ as follows:[21]

$$n_o = n_{\perp} \, , \tag{13}$$

$$n_e = \frac{n_{\parallel} n_{\perp}}{\sqrt{n_{\parallel}^2 \cos^2\zeta + n_{\perp}^2 \sin^2\zeta}} \, . \tag{14}$$

As the two split beams experience two different refractive indices they travel with different velocities through the medium. When coming out of the medium the two beams recombine but with a phase shift δ of:[21]

$$\delta = \frac{2\pi}{\lambda} \; (n_e - n_o) d \; . \tag{15}$$

In Equation (15) λ is the vacuum wavelength of the incoming light, d is the geometrical path of the light in the medium, and the difference between n_e and n_o is defined as birefringence $\Delta n = n_e - n_o$. Furthermore, it should be mentioned that the refractive indices, and therefore also the birefringence, are wavelength dependent. This phenomenon in optics is called dispersion and describes that the phase velocity of a wave depends on its frequency, or alternatively on its wavelength. The refractive indices of most transparent materials (like air, glasses) and also of LCs decrease with increasing wavelength according to $1 < n \; (\lambda_{red}) < n \; (\lambda_{yellow}) < n \; (\lambda_{blue})$, then the medium is called to have normal dispersion, whereas an increasing refractive index would correspond to anomalous dispersion.[21]

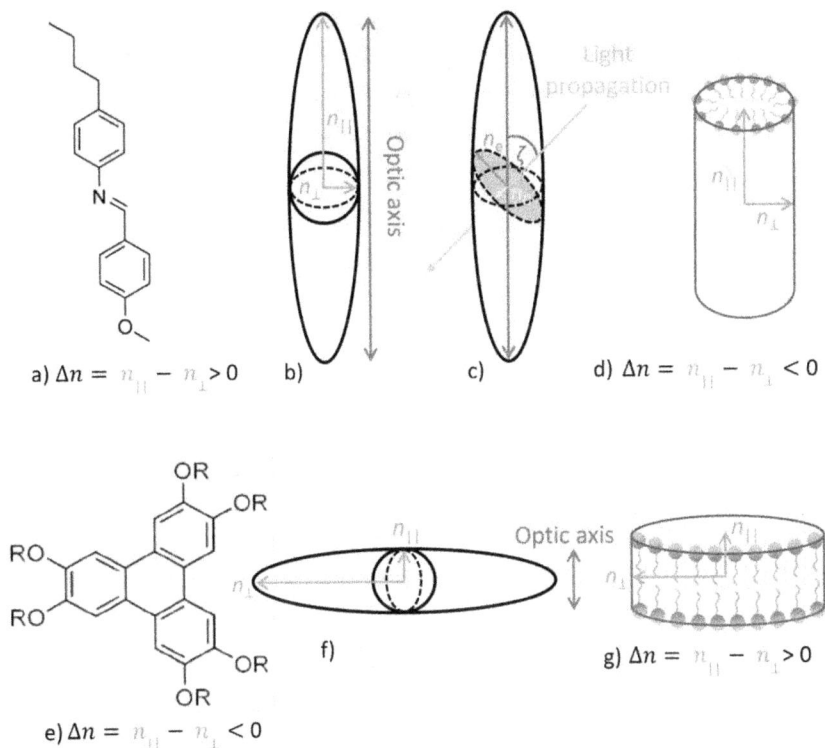

a) $\Delta n = n_{||} - n_{\perp} > 0$ b) c) d) $\Delta n = n_{||} - n_{\perp} < 0$

e) $\Delta n = n_{||} - n_{\perp} < 0$ f) g) $\Delta n = n_{||} - n_{\perp} > 0$

Figure 3.6: Schematic overview of the optical anisotropy in thermotropic and lyotropic liquid crystals: (a) Molecular rod-shaped structure of the thermotropic liquid crystal n-(4-methoxybenzylidene)-4-butylaniline "MBBA" being the building block for an optically positive N$_C$ phase. (b) Corresponding indicatrix of an optically positive uniaxial material with the two refractive indices $n_{||}$ and $n_{\perp}$. (c) When light propagates through the optically anisotropic medium at an angle ζ with respect to the optical axis, the incoming light beam is split into an ordinary beam (n_o) and an extraordinary beam (n_e). (d) Rod-shaped micelle of a lyotropic liquid crystal being the building block of an optically negative N$_C$ phase. (e) Molecular disc-shaped structure of a triphenylene being the building block for a thermotropic optically negative N$_D$ phase. (f) Corresponding indicatrix of an optically negative uniaxial material. (g) Disc-shaped micelle of a lyotropic liquid crystal being the building block of an optically positive N$_D$ phase.

Birefringence gives rise to the ability of a material to change the polarization state of light. Liquid Crystals are such materials and a powerful tool to identify and characterize liquid crystalline structures and textures is by polarizing optical microscopy (POM).[21]

The basic features of a POM are two polarizers, one located along the light beam before the sample and one after the sample. This second polarizer is also often called analyzer whereas the first is simply called polarizer. Typically the two polarizers are set perpendicular to each other at a 90° angle, this setup is called crossed polarizers.

A schematic figure of how a birefringent material between crossed polarizers affects the polarization state of light is shown in Figure 3.7. Unpolarised light is emitted by a lamp, hits the first polarizer and becomes linearly polarized oscillating at the same angle as the polarizer was placed. When the linearly polarized light beam enters the birefringent material it splits up into the ordinary and the extraordinary beams, each of them propagating with different speeds c/n_o and c/n_e through the sample, with c being the speed of light in vacuum. After exiting the birefringent material the ordinary and extraordinary beams recombine with a phase shift. If the phase shift is $\delta = n\,2\pi$ (with $n = 0, 1, 2,...$) the polarization state does not change, whereas if $\delta \neq n\,2\pi$ the polarization state of the resulting light beam differs from the initial state of the light before entering the birefringent material.[21] In Figure 3.7 the case of $\delta = \pi$ is shown. The liquid crystalline sample rotates the plane of polarisation around 90° such that the light is fully transmitted through the analyzer.

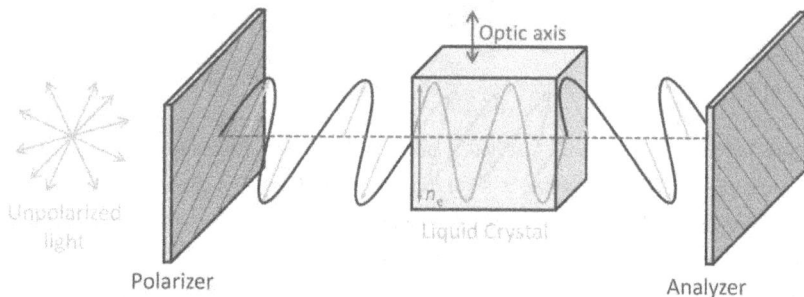

Figure 3.7: Schematic illustration of the light propagation through a birefringent liquid crystalline sample between crossed polarizers. The first polarizer linearly polarizes the unpolarized light emitted by the lamp of the microscope. Traveling through a birefringent sample the incoming light beam is split into the ordinary and the extraordinary ray which experiences two different refractive indices (n_o and n_e). The two beams travel at different velocities within the medium. After exiting the sample, the two beams recombine – in this figure exemplarily demonstrated – with a phase shift of $\delta = \pi$. This enables the light to pass through the second polarizer, the analyzer, which is set perpendicular to the first polarizer. The sketch is redrawn based on ref.[78].

The transmitted light intensity for the general case of crossed polarizers with any angle φ between the incoming plane of polarization and the optic axis of the sample is given as follows:[21]

$$I = I_0 \sin^2(2\varphi) \sin^2\frac{\delta}{2} \, . \tag{16}$$

I_0 is the light intensity after passing the first polarizer. If $\delta \neq 0$ the intensity is influenced by the birefringence and the sample thickness, see Equation (15). Equation (16) gives a maximum intensity for $\varphi = 45°$, whereas a minimum intensity is obtained at $\varphi = 0°$ and $\varphi = 90°$. For the case of minimum intensity, the optic axis is parallel or perpendicular to the polarization plane of the two polarizers. Therefore, the linearly polarized light experiences either only n_e or n_o, so the light beam is not split up into two components meaning that no phase difference can occur. There is one other option for how a liquid crystalline sample could appear dark under crossed polarizers. This is when the polarized light propagates along the optic axis which is equivalent to the interaction of light with an optically isotropic medium. In literature, this case is often referred to as pseudo-isotropic. In uniaxial liquid crystals, the pseudo-isotropic state can be observed for homeotropic alignment.[21]

The polarized optical micrographs of liquid crystalline samples observed under crossed polarizers in a POM with white light are typically very colorful. This beautiful phenomenon is due to the fact that the phase shift δ changes with wavelength, giving lots of interference colors. Auguste Michel-Lévy established an interference color chart, the so-called Michel-Lévy chart. This chart assigns the order of interference color with the birefringence, the sample thickness and the optical path difference, see Figure 3.8a.

Figure 3.8b – d demonstrate exemplarily how rotating the polarizers and adding a wave plate can help to figure out the director field in liquid crystalline textures. The example shown here is the escaped radial configuration with alternating ±1 point defects. The point defects mark the spots where the escape direction is changing. In Figure 3.8 the ERPD configuration is observed between crossed polarizers. It shows basically white, grey and yellowish colors. The color change from the middle of the capillary to the outer parts is due to the change in sample thickness (capillary diameter = 700 µm). Additionally, a fine dark line along the capillary axis and, at the point defects, also a blurry dark line perpendicular to it, can be seen. As mentioned above, these dark regions indicate that the local optic axis is parallel or perpendicular to the axes of the polarizers. Nevertheless, the director field in the other parts of the capillary is still not known.

When rotating both polarizers to an angle of ±45° with respect to their initial state (Figure 3.8c) this gives a darkening of the regions in which the director is tilted around ±45°, because then these are the regions where the planes of polarization of the polarizers are parallel or perpendicular to the local optic axis. For further investigations to determine the direction of the optic axis the use of a wave plate is helpful, see Figure 3.8d. Waveplates are made of birefringent materials and add a specific optical path difference along a uniform direction. This means, that the phase shift is increased along the optic axis of the wave plate which defines that direction.

In this thesis, a so-called λ plate for a path difference of $\lambda \approx 530$ nm is used. When the λ plate is inserted at an angle of 45° between the sample and the analyzer in the case of crossed polarizers, a color shift is induced, see Figure 3.8d. The black background and brushes become magenta. Looking at the Michel-Lévy chart reveals that the color change can be correlated to the first order magenta, which is indicated with a red dot and a pink arrow in Figure 3.8a. Furthermore, the initial grey regions become alternating blue and orange, which is due to the different orientation of the optic axis. The phase shift of the λ plate is either effectively added or subtracted to the optical path difference depending on whether the optic axis of the liquid crystal is aligned parallel or perpendicular to the optic axis of the wave plate. In the case of addition (+λ), a color shift to blue is observed, because coming from the white region in the Michel-Lévy chart a distance of $\lambda = 530$ nm coincides with the blue region. Whereas in the case of subtraction (-λ) the color changes to orange. To understand the subtraction, it has to be mentioned, that the Michel-Lévy chart extends to the left side as a mirrored image, where an axis of negative birefringence exists. This is where the -λ case is originally found. Consequently, the observations with a λ plate reveal the local orientation of the optic axis, which eventually corresponds to the director field. A λ plate is, therefore, a powerful tool for the investigation of liquid crystalline textures.

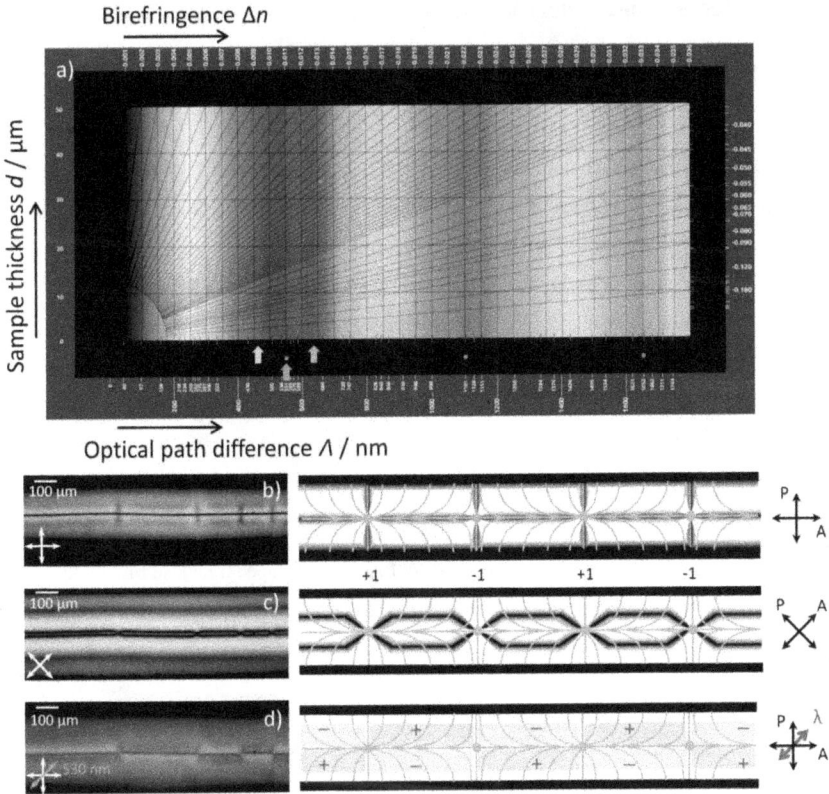

Figure 3.8: Exemplarily demonstration of the optical phase shift in birefringent materials. The here used material is the N_D phase of the lyotropic LC which is studied in the following. (a) Michel-Lévy chart showing the order of interference color depending on the birefringence Δn and sample thickness d, as well as their combined effect, the optical path difference Λ, reprinted from the Zeiss website.[79] (b) – (d) polarized optical micrographs on the left side and corresponding director field on the right side of the escaped radial configuration with alternating ±1 point defects separating regions of opposite escape direction. In (b) the ERPD configuration is observed between crossed polarizers, whereas in (c) the polarizers are rotated about ±45° from their initial state and in (d) by inserting a λ plate an additional phase shift towards blue or orange reveals the orientation of the optic axis. The combination of (b) – (d) allows a mapping of the director field.

3.3 Materials and experimental methods

3.3.1 Sample preparation

The lyotropic liquid crystal (LLC) used in this study is a ternary system containing N, N-dimethyl-N-ethylhexadecylammonium bromide (CDEAB, see Figure 1.10a) as the surfactant, decan-1-ol (DOH) as cosurfactant, and doubly distilled water as the solvent. CDEAB and DOH were purchased from Merck KGaA and were used without any further purification. First, CDEAB was dissolved in water, then the cosurfactant DOH was added. The mixture was stored in glass vials with a screw plug and sealed with parafilm to prevent solvent evaporation. In order to homogenize the mixture, the sample was stirred in a thermoshaker (Biosan, PST-60HL) at 40° and put on a roller (Phoenix Instruments, RS-TR05). Overall, the homogenization took 5 days with alternating use of the thermoshaker and the roller.

The phase diagram and the micelle dimensions were studied in detail by Görgens.[80] The binary mixture of CDEAB/H₂O forms above a certain concentration rod-like micelles, whereas when adding DOH as cosurfactant the ternary mixture forms disc-shaped micelles. Figure 3.9 shows the phase diagram of the ternary system CDEAB/DOH/H₂O for a constant weight ratio of m(CDEAB)/m(DOH) = 6.6.[80]

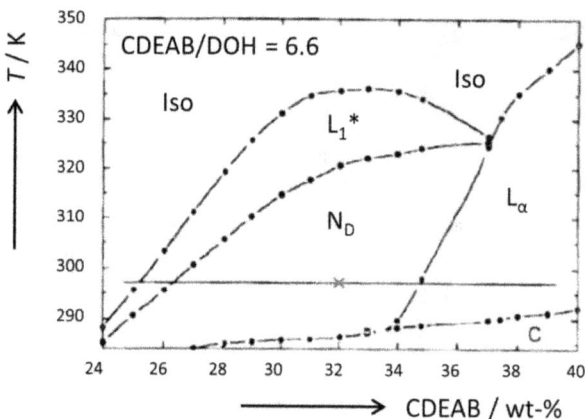

Figure 3.9: Phase diagram of the ternary system CDEAB/DOH/H₂O for a constant weight ratio of m(CDEAB)/m(DOH) = 6.6. The composition used in this study is marked with a red cross. Symbols: Iso = isotropic phase; C = crystalline phase; N_D = nematic phase with disc-shaped micelles; L_α = lamellar phase; L_1^* = optic isotropic phase with shear-induced birefringence. The diagram is reprinted with permission from ref. [80].

The phase diagram reveals a broad N_D phase at room temperature. In this thesis, the N_D phase formed at a composition of 32.0 wt% CDEAB, 4.8 wt% DOH and 63.2 wt% water was studied. The micelle dimensions at this composition were measured with light scattering by Görgens who reported a height of 32 Å and a diameter of 69 Å, excluding the solvation sphere. Taking the solvation sphere into account, the height is around 61.5 Å and the diameter around 80 Å.[80]

For the experiments on chiral induction, (R)- and (S)-mandelic acid (abbreviations: (R)-MA and (S)-MA) were purchased from Alfa Aesar with >99 % purity, >99 % enantiomeric excess and used as chiral dopants. Note that the calculation of the molar fraction of the added mandelic acid includes the water content of the system.

Mark capillaries made of glass no. 14 from Hilgenberg with an outer diameter of 0.7 mm and a wall thickness of 0.01 mm were used as capillaries. In supplementary studies, borosilicate glass capillaries from Vitrocom, like they were used in ref. [61], with an inner diameter of 0.15 mm were applied as well. The lyotropic mixture was sucked into the capillary in the nematic phase by using a water pump jet. After the capillary had been filled, both ends were sealed by melting the glass with a lighter. In order to check the sealing, the capillaries were centrifuged.

It is well known, that disc-shaped LLCs align homeotropically at the glass surface because the polar headgroups of the surfactant molecules interact with the polar Si-O- and Si-OH-bonds at the glass surface, see Figure 3.10.

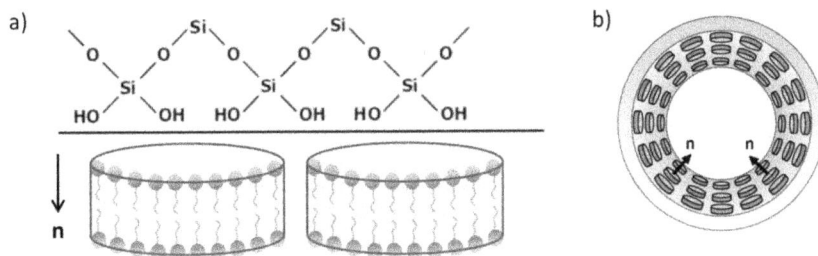

Figure 3.10: Homeotropic alignment of disc-shaped micelles. (a) The interaction between the polar Si-O- and Si-OH-bonds of the glass surface and the polar head groups of the surfactant molecules is largest if the disc orients perpendicular to the diameter of the micelle. (b) Schematic illustration of the homeotropic anchoring at the inner capillary glass walls; see ref. [45], reprinted with permission from Langmuir.

3.3.2 Alignment in a magnetic field

By means of applying a magnetic field to a liquid crystalline sample, the macroscopic orientation of the LC can be influenced. In lyotropic LCs, the amphiphilic surfactant molecules exhibit diamagnetic behaviour in a magnetic field. The resulting magnetic moment **M** is proportional to the applied field strength **H**, with the diamagnetic susceptibility χ_m being the material tensor:[5]

$$\mathbf{M} = \chi_m \cdot \mathbf{H} \ . \tag{17}$$

Because of the anisotropic conformations of the C-C- and C-H-bonds of the surfactant molecules, the anisotropy of the electron probability density is determined by the molecular structure which causes an anisotropy of the diamagnetic susceptibility. This molecular anisotropy is enhanced by the aggregation process forming micelles and lyotropic liquid crystals. It is this macroscopic anisotropy which evokes the macroscopic alignment effect of a magnetic field on a liquid crystalline sample.

As already mentioned in the Introduction, disc-shaped building blocks of amphiphilic molecules have a negative magnetic susceptibility anisotropy, meaning that a perpendicular alignment of the director towards the applied magnetic field is energetically favored.[5] Additionally, the use of an aligning magnetic field can facilitate the detection of helical superstructures, as the helix axis of cholesteric LCs with negative diamagnetic anisotropy is formed parallel to the magnetic field. Several studies on lyotropic-nematic and lyotropic-cholesteric LCs show that it is not the magnetic field strength which affects primarily the quality of the sample alignment; it is rather the retention time of the sample in the magnetic field.[80] This could be due to the very small diamagnetic susceptibility in LLCs, being in the order of $\chi_m = 10^{-11} \ \mathrm{m^3 kg^{-1}}$.[6]

A Bruker B-E 25v electromagnet with a maximum magnetic field strength of 1 T was used for the experiments with a magnetic field. After the capillaries were filled, they were put into a homemade temperature-controlled sample holder which was placed in the electromagnet such that the capillary axes are parallel to the magnetic field. The samples were first heated into the isotropic phase ($\approx$ 50°C) in order to rule out any possible effects of shear alignment from the filling process. Then, they were slowly cooled down (with a rate of 0.2 K/h) from the isotropic phase to the nematic phase at room temperature with the magnetic field applied along the capillary axes. Roughly speaking, the capillaries had been placed for about 1 – 2 week in the magnetic field.

3.3.3 Polarizing optical microscopy

Polarizing optical microscopy (POM) is the oldest and one of the most important methods used to study liquid crystalline samples in order to identify the individual phases as well as their respective phase transition temperatures and to investigate their textures.[21]

The polarizing optical microscope Leica DM-LP, in combination with an INSTEC TS62 hot stage and a Nikon D40 camera, was used for the studies in this thesis. The basic principle of POM is explained in Chapter 3.2. Using capillaries as confinement for the LLC required a special sample holder in order to be able to rotate the capillary while looking through the microscope. Figure 3.11 shows such a sample holder, which was manufactured by the workshop of the Institute of Physical Chemistry. The sample holder was placed on the hot stage of the microscope.

Figure 3.11: Photograph of the sample holder for the capillaries. With the small wheel on the left, the capillaries could be rotated smoothly.

Another obstacle was the lens effect which occurred due to the curved surface of the capillary glass, meaning that the outer parts of the capillary appeared black. To overcome that problem, the use of an index matching fluid, surrounding the capillary, was tested. In the case of the examples given in Figure 3.12a,b the use of rapeseed oil as index matching fluid (IMF) gave no significant additional information. Therefore, photographs in the first part of the results and discussion section of this chapter were made without an index matching fluid for reasons of simplicity unless otherwise stated.

However, as shown in Figure 3.12c the use of water as IMF provided significant additional information of the texture appearing under crossed polarizers. Furthermore, it seemed to be better to match the refractive index of the inside (LLC) than of the outside (glass) in case of curved surfaces confining the LC. Therefore, the use of water as IMF was more suitable than the use of some oil, because the refractive index of water is close to the one of the lyotropic liquid crystal, which consists of 63 % of water. Due to that, photographs in the second part of the results and discussion section of this chapter were taken with water as IMF surrounding the capillary.

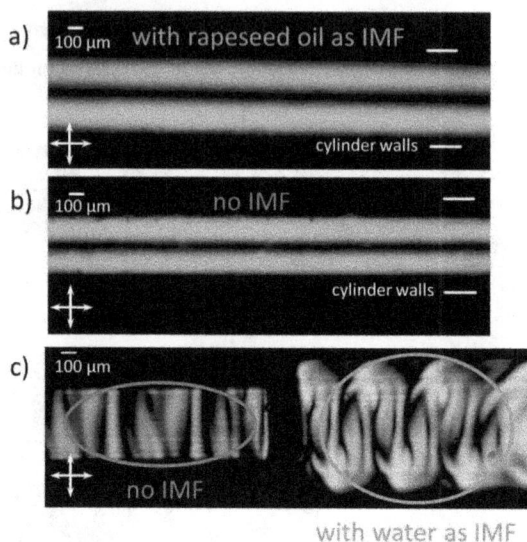

Figure 3.12: Illustration of the effects of an index matching fluid (IMF) surrounding the capillary. (a) and (b) POM images of the same LLC-filled capillary ($\emptyset = 0.7$ mm) surrounded by rapeseed oil as IMF (a) and not surrounded by an IMF (b). No significant additional information at the outer parts near the capillary walls can be obtained. (c) POM image of one capillary where the left part is not covered with an IMF and the right part is surrounded by water as IMF. Significant additional information on the observed texture is provided by the use of water as IMF. Figure from ref. [45], reprinted with permission from Langmuir.

3.4 Results and Discussion

3.4.1 Chiral configurations under capillary confinement

3.4.1.1 Twisted Escaped Radial Configuration (H = 0)

The first chiral configuration found in the achiral micellar nematic lyotropic system of CDEAB/DOH/H_2O is the twisted escaped radial (TER) configuration which appeared in the absence of an external magnetic field. In literature, this TER configuration was observed the first time in the case of chromonic LLCs and studied in detail by Jeong et al.[61].

The development of the director field with time without applying an external magnetic field is summarized in Figure 3.13. Capillaries that were filled with the nematic phase of the LLC showed first a typical schlieren-like texture with no macroscopic director alignment; see Figure 3.13a. After 2-3 days at room temperature, this schlieren texture transformed into a defect-free configuration as shown in Figure 3.13b which can be distinguished from the following picture in Figure 3.13c by the bright domains in the middle of the capillary. Waiting for an additional 2 weeks at room temperature these bright domains vanished and the brush in the middle of the capillary turned completely dark, see Figure 3.13c.

The director configuration observed in Figure 3.13b is the twisted escaped radial (TER) configuration, schematically drawn in Figure 3.2a. The configuration in Figure 3.13c is the non-chiral analog, the escaped radial (ER) configuration; its director field is schematically shown in Figure 3.1c.

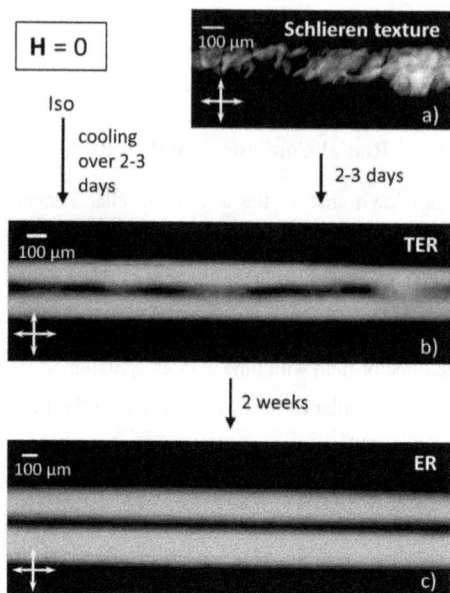

Figure 3.13: Time-dependent development of the LLC director field under capillary confinement in the absence of a magnetic field. In (a) the typical nematic schlieren-like texture is shown which was observed directly after filling. (b) After 2-3 days waiting at room temperature or slowly cooling down from the isotropic to the nematic phase a twisted escaped radial (TER) configuration with sections of alternating twist is obtained. (c) After waiting an additional 2 weeks at room temperature the bright domains disappear and the director field has transformed into a unidirectional escaped radial (ER) configuration which constitutes the ground state. Figure from ref. [45], reprinted with permission from Langmuir.

The TER configuration is characterized by a director twist along the diameter of the capillary. In the Mauguin limit, where the total twist angle is much smaller than the retardation caused by the anisotropy of the refractive indices, the director field serves as a waveguide to the incident polarized light.[81] This results in a brightening of the center part of the capillary under crossed polarizers in the TER due to – in case of a Mauguin number greater than 1 – a rotation of the plane of polarization in addition to the retardation of traveling light.[64,81,82] Because the birefringence of LLCs is, in general, two orders of magnitude smaller than the birefringence in thermotropic LCs, a rough estimation reveals a Mauguin number of ≥ 1.[6]

The bright domains are separated without a defect in between by twist-free regions which appear black due to extinction under crossed polarizers, see Figure 3.14a. Since the lyotropic liquid crystal is inherently non-chiral, domains of both senses of handedness should be formed with the same probability under spontaneous reflection symmetry breaking. The local sign of twist sense in the TER configuration can be easily verified by a slight decrossing of the polarizers; see Figure 3.14b,c,e,f.

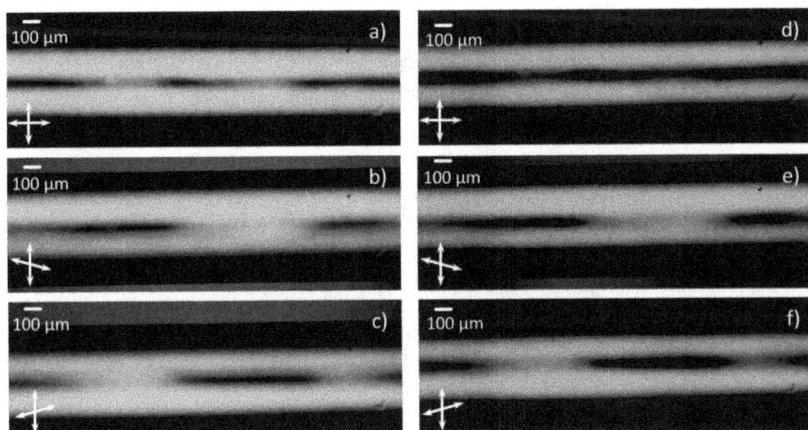

Figure 3.14: Confirmation and determination of the handedness in twisted escaped radial domains. Regions of opposite twist sense appear with the same probability due to the fact that the LLC is non-chiral. (a) – (c) show polarized optical micrographs under white light illumination, (d) – (f) show the same under green light. (a) TER domains appear bright and orange under crossed polarizers and white light illumination. (b) and (c) shows that a slight decrossing of the polarizers (b) clockwise and (c) counterclockwise brightens or darkens regions of opposite twist sense. Figure from ref. [45], reprinted with permission from Langmuir.

As illustrated in Figure 3.13, after waiting for some weeks, the TER configuration relaxes into the non-chiral ER configuration. This indicates that the state of lowest energy, the ground state – at least at room temperature, is not the twisted escaped radial, but simply the ER which has reflection symmetry and is therefore non-chiral.

Of course, now the question arises why the relaxation time is so long. There could be several reasons like e.g. the characteristic lengths are very large since the diameter of the capillary is nearly one millimeter, and the wavelength of deformations along the axis could be some millimeters too, which in itself should give a slow dynamic behavior.

The initial schlieren texture which contained lots of topological defects transformed also very slowly to the TER configuration. This very slow transition from TER to ER suggests that the energy difference between those two configurations is very small. And like it was discussed by Jeong et al. in the case of chromonic LCs,[61] the occurrence of the TER configuration can be related to the relative magnitudes of the elastic constants meaning that also in standard micellar LLCs the twist elastic constant could be one order of magnitude smaller than the bend and splay constants.

Furthermore, the sensitive energy balance between TER and ER could be influenced by external stimuli like temperature, the composition of the ternary system giving different micelle dimensions and/or flow effects. One speculation could be that if the composition of the system were closer to the lamellar phase, the bend and twist elastic constants would be larger in comparison to the splay constant, giving also a reason why twist would be suppressed and the formation of the non-chiral ER configuration would be favored. On the contrary, at lower CDEAB concentrations, further away from the lamellar phase in the phase diagram (Figure 3.9), the twist elastic constant could be very small permitting the director field to relieve splay and bend energy by adding a twist and giving reflection-symmetry broken configurations like the TER. Additionally, flow along the capillary axis would stabilize the ER over the TER configuration.

In conclusion, comparing the relaxation times of LLCs with thermotropic LCs the following rough estimation can be done: In thermotropic LCs, characteristic deformation wavelengths of ≈ 5 μm result in relaxation times τ of ≈ 5 ms.[14] The characteristic lengths in the TER configuration in the LLC are in the millimeter range (≈ 1 mm). Without considering a difference in elastic constants of thermotropic and lyotropic LCs, the relaxation time would be in the order of hours according to the simple assumption of $\tau \sim length^2$.[14,83] Whereas when taking a difference in the elastic constants into account, assuming that the elastic constants of LLCs are one order of magnitude smaller than those in thermotropic LCs, a relaxation time in the order of days would be obtained.

3.4.1.2 Twisted Polar Configuration (H > H$_c$)

The twisted escaped radial configuration is not the only chiral configuration that was found in the achiral lyotropic system of CDEAB/DOH/H$_2$O under capillary confinement. The second chiral configuration which appeared – however only when an external magnetic field was applied – is the so-called twisted polar (TP) configuration which is characterized by two half-unit disclination lines twisting around each other and forming a double helix along the capillary axis. This configuration has been observed in the case of capillary-confined chromonic LLCs as well. However, as mentioned in the introduction of this chapter (see Figure 3.4 and Figure 3.5) Jeong et al.[61] named this configuration twisted planar polar (TPP) but have not studied it as intensely as they did in the case of the TER, e.g. no further explanation concerning the formation process of this TPP was provided in ref. [61]. In the following, this chapter will give a more detailed interpretation of this configuration and will demonstrate that it should be named twisted polar configuration (TP).

Before showing the TP configuration, let us have a closer look at what happens to the known ER configuration when an axial magnetic field is applied at room temperature. In Figure 3.15a the quiescent state of the ER configuration observed between crossed polarizers is shown. When applying a magnetic field axial to a capillary filled with the LLC, the director of the N$_D$ phase orients perpendicular to the direction of the magnetic field due to the negative magnetic susceptibility anisotropy. The director is squeezed into a plane normal to the direction of the magnetic field, as it is schematically drawn in Figure 3.15b.

When removing the sample from the magnetic field an image of a transient field distorted escaped radial configuration, as shown in Figure 3.15c, is observed in the POM. Three extinction bands running parallel to the capillary axis can be seen. However, having a closer look, only the fine line in the middle is completely dark, whereas the symmetrically positioned neighboring bands are rather broad and greyish, giving no perfect extinction. In the dark line around the capillary axis, the director is parallel to this axis and forms a "stabilized kink" which gives this fine extinction region between crossed polarizers. Away from the capillary axis towards the cylinder wall, the director field is distorted by splay and bend deformations. In two small regions symmetrically just above and below the fine black extinction line, the director differs from a parallel or normal orientation along the capillary axis. These regions, where the director is tilted maximal 45° with respect to the capillary axis appear bright because the capillary axis is parallel to the polarizer. Passing these regions, at some radial distance, the director gradually orients normal to the cylinder axis (while the capillary thickness gradually decreases simultaneously) meeting the homeotropic boundary conditions at the capillary walls and appearing black between crossed polarizers. This field distorted escaped radial configuration is only transient because within 6 h the system relaxes completely to the quiescent escaped radial configuration. Additionally, the distorted ER is observed only when the capillaries were treated with an external magnetic field which confirms that its occurrence is only facilitated by the action of an axial magnetic field.

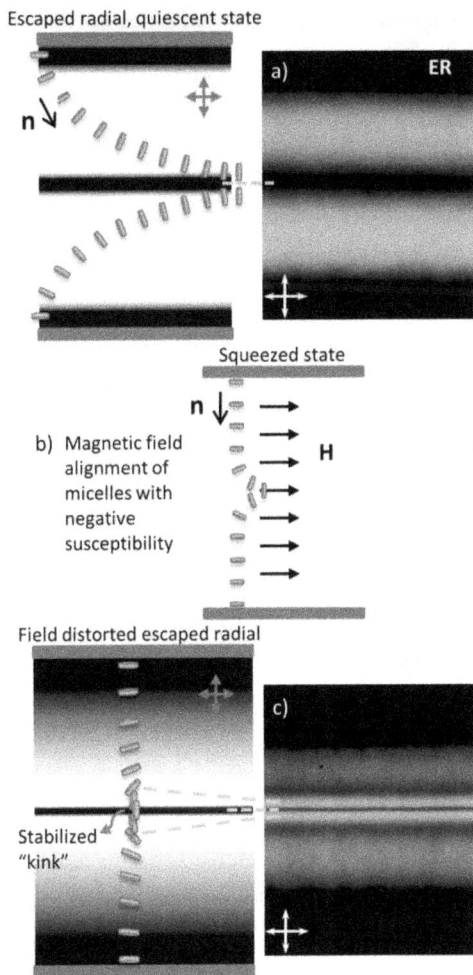

Figure 3.15: Schematic illustration of what happens to the director field **n** of the escaped radial configuration under a magnetic field applied along the axis of the capillary. On the right side, the images observed under the POM and on the left side the structure and optical features are illustrated. a) Shows the ER configuration between crossed polarizers. b) A magnetic field squeezes the ER configuration into a more planar configuration due to the negative anisotropy of magnetic susceptibility of the disk-shaped micelles of the LLC which are represented as blue discs. c) After the magnetic field is removed, a distorted escaped radial state with three dark band regions is observed for several hours until it relaxes to the initial non-distorted state of the ER configuration. Figure from ref. [45], reprinted with permission from Langmuir.

If the magnetic field is applied along the capillary axis during very slow cooling from the isotropic to the nematic phase, a more complex configuration was observed, shown in Figure 3.16. The LLC CDEAB/DOH/H_2O filled into a 700 µm thick capillary has been cooled down with a rate of 0.2 K/h from the isotropic phase to the N_D phase at room temperature under an axial magnetic field of 1 T and then remained additional 2-3 days in the magnetic field at 25 °C. The observed texture is shown in Figure 3.16a which depicts an approximately 23 mm long section of the capillary. Typically, the capillaries were 50 mm long. Starting at the left end of Figure 3.16a which is shown enlarged in Figure 3.16b, three extinction bands running parallel to the capillary axis can be observed. However, having a closer look at the magnification shown in Figure 3.16b, only the fine line in the middle is completely dark, whereas the symmetrically positioned neighboring bands are rather broader and greyish, giving no perfect extinction. This is the previously mentioned field distorted escaped radial configuration, explained in Figure 3.15.

Further along the capillary in Figure 3.16a, there is a transition to a configuration with two disclination lines emerging from a point on the cylinder axis. Proceeding to the right, these disclination lines form a double helix with a period (pitch) of roughly 3.5 millimeters. Additionally, the handedness of the double helix is changing several times moving from the left to the right in the photograph in Figure 3.16a. First, it changes after 1.5 turns, then after one full turn and then again after 1.5 turns. Even though the number of full turns is few, it appears that the pitch is the same for both senses of handedness.

In Figure 3.16c,d enlarged sections of the double helix with opposite twist senses are shown. The handedness of the double helix can be easily determined by rotating the capillary about its axis and checking if the crossover points of the two disclination arms are moving to the left or to the right, see Figure 3.16i,j and supporting information movies S1 and S2. In Figure 3.16c the double helix is left-handed and in Figure 3.16d it is right-handed. Between a right-handed double helix and a left-handed double helix, the twist sense is changing continuously. When rotating the capillary about its axis in such a region of change in handedness, the double helix arms move towards each other.

As the formation of a field distorted escaped radial configuration was already discussed in Figure 3.15, let us continue with the discussion of possible planar radial (PR) and planar polar (PP) configurations and the branching of one +1 defect into two + ½ defect lines at a transition point. As a magnetic field is applied along the capillary axis a continuous distortion of the ER configuration into a more planar configuration was observed, see Figure 3.15c.

Above some critical H_c, the total free energy of a planar radial (PR) with one +1 defect line or a planar polar (PP) configuration with two + ½ disclination lines should become smaller than that of a field distorted ER configuration. Therefore, one could suggest a field-induced transition from the squeezed field-distorted ER to the PR or PP configurations. After removing the magnetic field, the system should relax with time back into the ER ground state. However, this scenario is not what was observed.

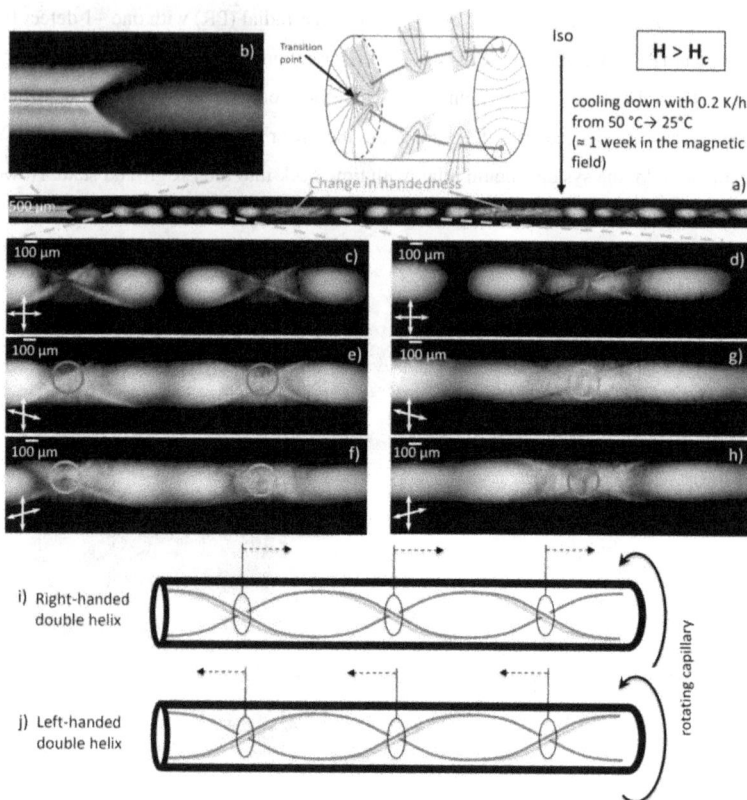

Figure 3.16: Twisted polar configuration observed in capillaries with $\emptyset = 700\,\mu m$. The LLC was cooled down from the isotropic phase to the nematic phase with a rate of 0.2 K/h in an axial magnetic field $H > H_c$ (1 Tesla). (a) Polarized optical micrograph of an approximately 23 mm long capillary section. (b) The transition point between the distorted escaped radial and twisted polar (TP) configuration. (c) – (d) Magnifications of regions with opposite twist sense of the TP double helix, in (c) the double helix is left-handed and in (d) it is right-handed. (e) – (h) At the encircled crossover points the twist along the diameter of the capillary (transverse twist) is normal to the plane of the paper. A clockwise and counterclockwise decrossing of the polarizers results in a color shift (orange to bluish or the opposite) depending on whether the axial twist of the double helix is left- or right-handed. This reveals an opposite sign of the transverse twist for left- and right-handed axial twist regions. (i) – (j) Determination of the handedness of the axial twist when rotating the capillary about its axis. Rotating the capillary clockwise the crossover points of a right-handed double helix move to the right, whereas they move to the left in case of a left-handed double helix. Figure from ref. [45], reprinted with permission from Langmuir.

Unfortunately, with the available equipment, it was not possible to observe the formation process of the TP configuration when the sample was inside the magnetic field. Therefore, the possible occurrence of a PR configuration which could be indeed formed at any time during the cooling from the isotropic to the nematic phase at **H** > **H**c, cannot be proven since the PR would relax immediately to the ER configuration, as soon as the field is removed, hence, before the capillary was put in the microscope.

Furthermore, the PP configuration with two $s = + \frac{1}{2}$ disclinations should be energetically favored upon the PR configuration with one $s = +1$ disclination, since the elastic energy stored around a disclination per its unit length, the so-called line tension, is proportional to s^2.[15] Therefore, a field-induced PR should be immediately transformed into the PP configuration which is lower in energy.

The branching of one +1 defect line (PR) into two + ½ defect lines (PP) is studied in detail theoretically by Shams et al.[84], however, the singularity-free ER configuration is not discussed in this context. In accordance with the discussion above, a transition from the field distorted squeezed ER to the PP configuration under an axial magnetic field could be expected, and when removing the magnetic field the induced PP would relax back to the quiescent ER. The point is, that if the PP would be metastable, a possible coexistence of the PP and ER configurations and the defect (de)branching should be observed under the microscope which is not the case.

Figure 3.16b however suggests that it is more of an abrupt transition from the field distorted ER to a metastable polar configuration stabilized by the action of a magnetic field containing a point defect and a planar radial configuration at the transition point (see right of Figure 3.16b), analogous to the disclination branching studied by Shams et al.[84].

Comparing the TPP configuration from ref. [61] with the observed structure in Figure 3.16c,d reveals the fact that the double helix does not appear dark between crossed polarizers along and perpendicular to the capillary axis which means that it´s not a planar structure. A slight decrossing of the polarizers indicates that the configuration contains not only the obvious axial twist along the capillary axis but also an additional transverse twist running across the two + ½ disclination lines parallel to the capillary diameter, as demonstrated in Figure 3.16e – h. Only in the middle between two crossover points extinction was observed because in these black domains the twist along the diameter of the capillary lays horizontal meaning parallel to the plane of the page.

In the case of the TER configuration, the conclusion could be drawn, that as for chromonic LLCs the twist elastic constant in the micellar LLC is relatively small compared to the splay and bend constants.[61] Due to that, the system would minimize its elastic free energy by reflection symmetry breaking. The scenario for the reflection symmetry breaking of the PP into the TP configuration is however not so easy to understand. The extremely weak axial twist with a periodicity in the millimeter range would not be able to lower the elastic energy compared to the non-twisted PP configuration. Quite the contrary, it would be reasonable if the system would favor the non-twisted PP configuration in order to minimize the length of the energetically costly disclination lines because the disclination lines would be shorter in the non-twisted case.

The reason for the reflection symmetry breaking however could lie in the disclinations themselves, as they could serve as chiral seeds. The director field around the disclination lines is strongly deformed and in agreement with the argument – like in the case of the TER – that strong bend and splay deformations can be minimized by adding twist deformation, if the twist elastic constant is much smaller than the splay and bend constants, this would lead to reflection symmetry broken configurations. This way of symmetry breaking is illustrated in Figure 3.17. The director field is represented by cylinders and the disclination lines are marked as red dots/lines. In Figure 3.17a the case of a + ½ wedge disclination is drawn. The director field is in a plane normal to the disclination line and contains solely splay and bend deformations.[85,86] This would be the corresponding director field of the PP configuration.

Nevertheless, as discussed by Frank,[20] a half unit disclination line can be also realized by rotating the director 180° out of the plane, as it is shown in Figure 3.17b,c. This is then called twist disclination.[85–87] The director follows the plane in which the molecules lie or in other words putting the director on a Möbius strip encircling the disclination.[85] The twist disclination can be obtained from the wedge disclination by a rotation of 90° about an axis vertical in the plane of the paper; the axis is marked as an orange arrow in Figure 3.17b. Depending on the oblique side view direction the twist disclination can be either + ½ or – ½, see Figure 3.17b. If the TP configuration is viewed from the angle indicated above the blue arrow, it is a $s = + ½$ twist disclination,[20] see Figure 3.17c,d. Due to the fact that in twist disclinations a director twist is introduced, the reflection symmetry is broken. According to Ranganath,[88] Anisimov, and Dzyaloshinskii,[89] in usual nematic LCs the non-chiral wedge disclination is energetically favored upon the chiral twist disclination, except a very small twist elastic constant K_2 comes into play with $K_2 < \frac{K_1+K_3}{2}$, K_1 being the splay and K_3 being the bend elastic constant.

In Figure 3.17e a cross-section of the TP director field under the capillary confinement is drawn schematically. The twist around the ½ disclinations has the opposite twist sense of the transverse twist which goes parallel to the diameter of the capillary connecting the two disclination lines. Viewing the scenario from the side, as shown in Figure 3.17f, the transverse twist evokes a non-parallel progression of the disclination lines, forming the double helix observed along the capillary axis (axial twist). As the transverse twist governs the sign of the axial twist, both have the same handedness which is the opposite to the twist sense around the disclinations.

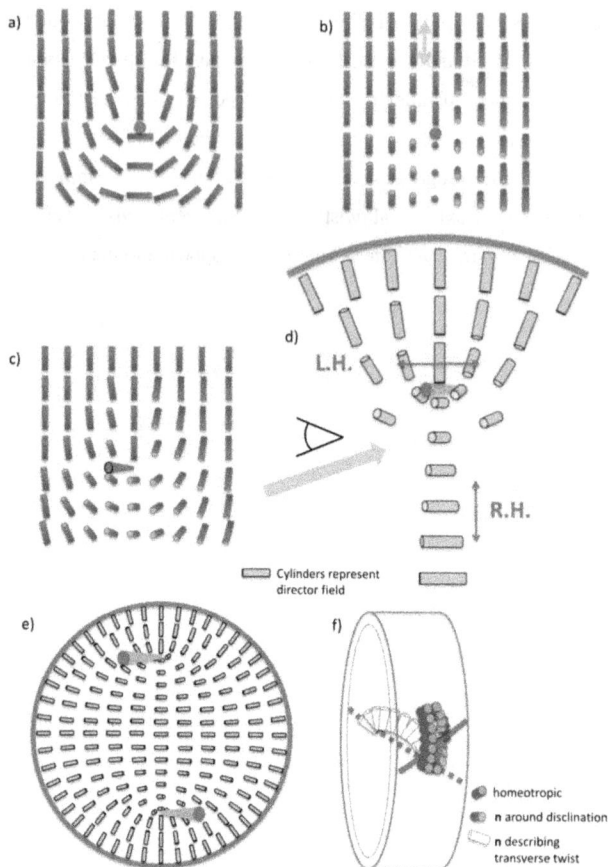

Figure 3.17: Schematic illustration of the reflection symmetry breaking in the vicinity of the + ½ disclination lines (marked as red dots/lines) in the twisted polar configuration. (a) Top view of the director field (here represented by cylinders) of an achiral + ½ wedge disclination. The director lies in a plane normal to the disclination line and contains only splay and bend deformations. (b) Top view of the director field of a chiral ± ½ twist disclination, the director twists out of the paper plane and around the disclination line following the plane in which the molecules lie. The twist disclination is obtained by rotating the director field from the wedge disclination by an angle of 90° about a vertical axis in the plane of the page (axis is indicated as an orange arrow in (b). (c) Oblique side view on a + ½ twist disclination. (d) – (f) two ½ twist disclinations transferred to the cylindrical confinement of a capillary. The director twist around the twist disclinations has the opposite handedness compared to the transverse and axial twists going along and perpendicular to the diameter of the capillary, respectively. Figure reprinted with permission from Langmuir from ref. [45].

Figure 3.18a shows exemplarily an entire capillary (4.6 cm in length) containing the TP configuration. The photo was taken between crossed polarizers without the magnification of a microscope. The + ½ disclination lines can be identified as the fine bright lines forming a double helix. In the middle of the capillary, an air bubble disrupts the continuous TP configuration. For reasons of comparison to the case of chromonic LLCs, capillaries with a diameter $\emptyset$ of 150 µm were tested as well. The TP configuration appeared just as in the capillaries with 700 µm diameter, see Figure 3.18b.

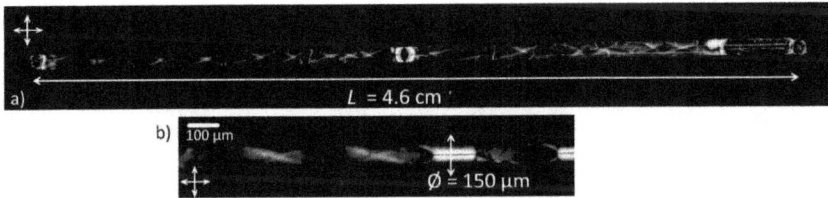

Figure 3.18: (a) Photograph of an entire capillary between crossed polarizers without the magnification of a microscope. The capillary has a length of approximately 4.6 cm and the continuous TP configuration is solely disrupted by an air bubble in the middle of the capillary. The fine bright lines forming a double helix are the + ½ disclination lines. (b) Polarized optical micrograph of a TP configuration in a capillary with a diameter of 150 µm between crossed polarizers.

Figure 3.18a shows also that the axial periodicity can vary throughout the capillary, compared to the relatively uniform periodicity of the TP configuration shown in Figure 3.16a. It was observed that the retention time of the capillary within the magnetic field affects the uniformity of the axial periodicity. The longer the capillaries were objected to the magnetic field (up to several weeks or even one month), the more regular the axial periodicity became.

As one could expect, the axial periodicity decreases with the diameter of the capillary. Whereas the axial periodicity was found to be in the range of 4 – 5 millimeters in $\emptyset$ = 700 µm capillaries, the axial periodicity in $\emptyset$ = 150 µm capillaries is around 1 mm.

An unsolved question is still remaining, namely, when the reflection symmetry breaking does take place. On the one hand, the twisting might occur directly in the branching process. The two disclination lines emerge in a rotating fashion like the two water beams in a garden sprinkler, while the sprinkler or the branching point is moving along the axis of the capillary. When the lines reached their equilibrium position near the capillary walls the double helix is formed.

The direction of rotation is arbitrary since it is still an achiral system and left- or right-handed twist sense can occur with the same probability. A small disruption in the formation process by dirt particles or point defects could impact the handedness of the TP configuration.

In the case of chromonic LCs, the formation of the double helix is performed by the emergence of two disclination lines from a moving point.[61] The TP configuration of this micellar LLC can be formed as well like that, but unfortunately, it cannot be tracked via POM due to the fact that it only forms under the action of a magnetic field. Another option would be that first a PP configuration is formed and then a twist is added. Figure 3.19 summarizes the energy discussion for the several director configurations, under the action of a magnetic field and with no magnetic field applied.

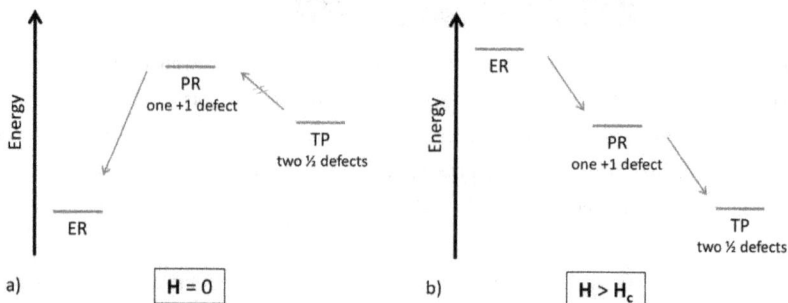

Figure 3.19: (a) Energy situation for the escaped radial (ER), the planar polar (PR) and the twisted polar (TP) configurations under capillary confinement without the action of a magnetic field **H**. (b) Energy situation for the same configurations under the action of a magnetic field **H** > **H**$_c$.

As Figure 3.19b illustrates, the TP configuration is lowest in energy when being exposed to an external magnetic field. The ER is unstable in the magnetic field, as the director is squeezed into a more planar configuration, as shown in Figure 3.15b. At some point, a planar radial configuration with one +1 defect is formed, but this is again unstable towards splitting into two ½ disclination lines, as the elastic energy stored around a disclination per unit length – the so-called line tension – is proportional to the square of the defect strength.[15]

When being outside of the magnetic field, the energy scenario is different, see Figure 3.19a. The ER is now the true ground state, but the TP formed under the action of a magnetic field cannot relax to the ER ground state via the energetically higher PR configuration, because this would mean that the two ½ disclinations in the TP would have to recombine into a +1 disclination. This is not possible because two ½ disclinations are always lower in energy than one +1 defect.[15] The TP state is, therefore, a state which is "arrested" by the presence of defects. What instead actually happens to the TP when being outside of a magnetic field for a longer time is discussed below.

3.4.1.3 Defect Stabilization of the Twisted Escaped Radial Configuration (TERPD)

In the previous chapter, several configurations that appeared under the action of a magnetic field were discussed. Now let´s have a look at what happens to those director fields when not subjected to a magnetic field for a longer time. It sometimes occurred that the periodicity of the axial twist of the TP configuration is very large up to a point where one might think it is a planar polar (PP) configuration, especially if only one-half turn can be seen; see Figure 3.18b on the right and Figure 3.20a. However, by rotating the capillary, no position in which the two disclination lines were completely parallel to each other – meaning the disclination line running in the back would be hidden by the one in the front – could be found, see Figure 3.20b.

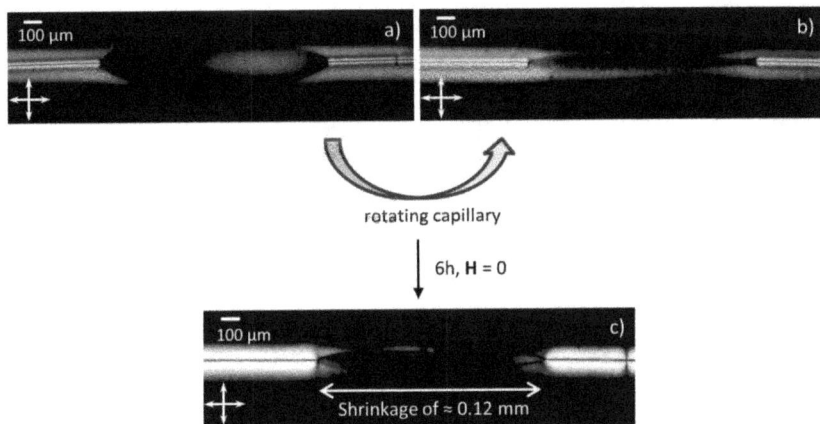

Figure 3.20: (a) Field distorted escaped radial configuration and one-half turn of a TP configuration with a very large periodicity of the axial twist observed directly after magnetic field treatment. (b) Rotating the capillary reveals that there is no position in which the disclination line in the front hides completely the disclination line running in the back confirming that this is a TP configuration with a very large periodicity of the axial twist. (c) After waiting for 6h with the capillary not been subjected to a magnetic field, a shrinkage of approximately 0.12 mm of the TP area was observed. The field distorted escaped radial configuration transforms into the quiescent state of the escaped radial configuration.

After waiting for 6h, a photograph of the same position was taken, see Figure 3.20c. The fine dark extinction line of the field distorted escaped radial configuration gets broader, meaning that it relaxes into the quiescent state of the escaped radial configuration. Measuring the length of the TP area from the transition point on the left to the transition point on the right, a decrease of approximately 0.12 mm was observed. This means that the transition points move with a speed of roughly 10 µm/h in order to enlarge the ER regions, which shows that the ER configuration is indeed the ground state.

However, if the area of the TP configuration is very large containing lots of turns of the double helix, the TP configuration relaxes in the absence of a magnetic field into a twisted escaped radial configuration with point defects (TERPD), see Figure 3.21. The TER configuration can be characterized, like in Chapter 3.4.1.1 by its bright appearance between crossed polarizers. A large number of crosses formed at the cylinder axis is interrupting the bright domains. These are point defects separating regions of opposite escape directions which is revealed by inserting a $\lambda = 530$ nm plate, see Figure 3.21d. Decrossing the polarizers indicates the TER twist along the diameter of the capillary and the presence of alternating +1 and -1 point defects, as shown in Figure 3.21b,c. Figure 3.21f gives a schematic illustration of the TERPD director field.

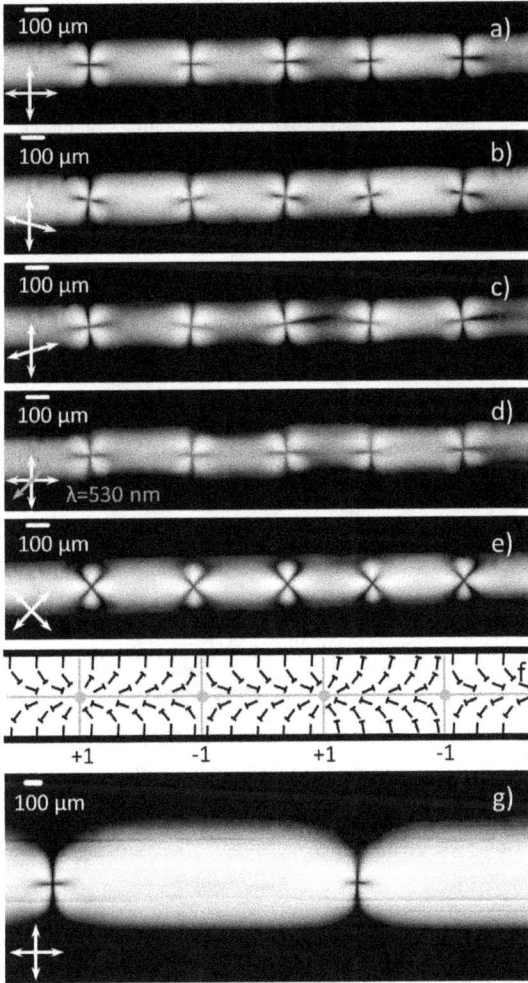

Figure 3.21: Defect-stabilized twisted escaped radial (TERPD) configuration obtained after relaxation from the twisted polar (TP) configuration by not being subjected to an external magnetic field. (a) The bright regions in the center indicate the twisted state, which is further verified by decrossing the polarizers in (b) – (c) and rotating the polarizers by 45° around the cylinder axis in (e). By inserting a $\lambda = 530$ nm plate in (d), it is revealed that the point defects separate regions of opposite escape directions. The pink arrow indicates the slow axis of the optical compensator inserted at 45° to the polarizers. (f) Schematic illustration of the TERPD director field. (g) Even after half a year, the TER configuration is stabilized by point defects. Figure reprinted with permission from Langmuir from ref. [45].

The transition from the TP to the TERPD configuration was not possible to observe because it was happening very slowly in the time range of several weeks up to 2 months. As long as the defects are present, the non-chiral ER configuration did not form and the chiral TERPD configuration is stable. Figure 3.21g shows that even after half a year the TERPD configuration is preserved. This configuration corresponds to a locked metastable state because the defect-free ER configuration would be the ground state. However, as discussed by Allender et al., the boundary conditions at the ends of the capillary prevent the director field to relax into the ER configuration.[90] As shown in Figure 3.20, in the regions of the capillary which do not have the TP configuration containing some double helix turns, the defect-stabilized TER configuration relaxes into the achiral ER.

The question of whether the handedness of the TER in the defect-stabilized state is related to the handedness of the former TP could not be answered due to the very long transition times. However, if this would be the case, the reflection symmetry breaking originating from the formation of the TP configuration would be also preserved in the TERPD configuration. Furthermore, for the same reasons as in the case of the half-unit twist disclinations, the point defects in the TERPD might form chiral seeds because of the strong elastic deformation in their vicinity, which spontaneously leads to twisting.

Due to the very slow time scales of the transitions between the numerous configurations (ER, TER, TP, and TERPD) observed under capillary confinement in the investigated micellar lyotropic LC, the energy difference between these configurations are very tiny and it´s not so simple to identify the energetic ground state. The global stability of a certain configuration critically depends on weak external or internal stimuli, like the presence of a magnetic field or the formation of point defects. For example, the TP configuration is stable under the action of a magnetic field of sufficient strength. However, after removal of the magnetic field, the defect-stabilized TER configuration is observed which seems to be a locked state that can no longer relax into the non-chiral ER ground state. As the N_D phase of the CDEAB/DOH/H_2O system already forms heterochiral structures, this piques the curiosity if the chiral configurations can be stabilized by adding a chiral dopant as a stimulus. Moreover, to investigate how homochirality is developed, whether there is a threshold, how the chiral induction takes place and how the chiral configurations are influenced in particular. This is discussed in the following section.

3.4.2 Regimes of chiral induction

3.4.2.1 Doubly Twisted Escaped Radial Configuration (H = 0)

Due to the fact that the micellar nematic LLC system of CDEAB/DOH/H_2O is achiral, all reflection symmetry broken configurations under capillary confinement which were discussed so far contained both – left- as well as right-handed – twist senses with the same probability because there is no energetic bias for one handedness in achiral systems. This degeneracy is lifted when adding an enantiomeric chiral dopant of one handedness, in this case (R)-mandelic acid. The pitch of the chiral nematic phase induced by the chiral dopant decreases with an increasing amount of added dopant, or in other words, the inverse of the pitch, the so-called twist, increases with increasing dopant concentration. The molecular chirality of the chiral dopant is transferred to the macroscopic nematic director field which reflects the added chirality in terms of twisting. Because the chiral configurations in the achiral LLC under capillary confinement already contain a twist, the delicate balance between left- and right-handed twist senses can easily be biased towards one handedness by a chiral additive resulting in an extreme sensitivity to chirality. In addition, using capillary confinement gives a geometric advantage to develop an unhindered helicity over centimeters along the long axis of the capillary adding also to a high sensitivity for chirality in this system.

The effect of (R)-mandelic acid ((R)-MA) in the concentration range of 6 mmol% - 90 mmol% on the cylindrical confined nematic LLC under the absence of a magnetic field is summarized in Figure 3.22. It has to be mentioned, that the photographs now – compared to those in the previous chapters – have been taken using water as an index matching fluid, otherwise it would not be possible to understand these structures, see Figure 3.12c. The photographs were taken at room temperature since the molar fractions of the dopant are so little that a change in the chiral nematic to isotropic phase transition temperature can be neglected, see phase diagram in ref. [91].

First, up to a concentration of ≈ 20 - 30 mmol%, a continuous homochiral twisted escaped radial configuration is observed, in clear contrast to the achiral case in Figure 3.13b where domains of opposite twist sense were separated by twist-free regions. Because it is known from literature[30,37,92,93] that (R)-MA induces a left-handed director twist in our nematic LLC host phase, a schematic illustration of a left-handed TER director field is shown in Figure 3.23a.

Increasing the dopant concentration further from 40 - 80 mmol%, a new structure was observed, see Figure 3.22c–f. This configuration is characterized by a spiraling of the escape direction around the capillary axis giving a sinusoidal appearance. The sinusoidal amplitude, as well as the frequency, is increased when increasing the dopant concentration and thus the chirality. When rotating the capillary, the structure moves to right/left depending on the rotating direction, which suggests a phone cord-like structure. A schematic explanation of a possible director field with the comparison to a phone cord is given in Figure 3.23. Because there is an additional twist along the axis of the capillary in terms of a spiraling brush, the configuration was named doubly twisted escaped radial (DTER) configuration. Increasing the dopant concentration to 90 mmol% and further, henceforth the well-known fingerprint texture in Figure 3.22g was observed.

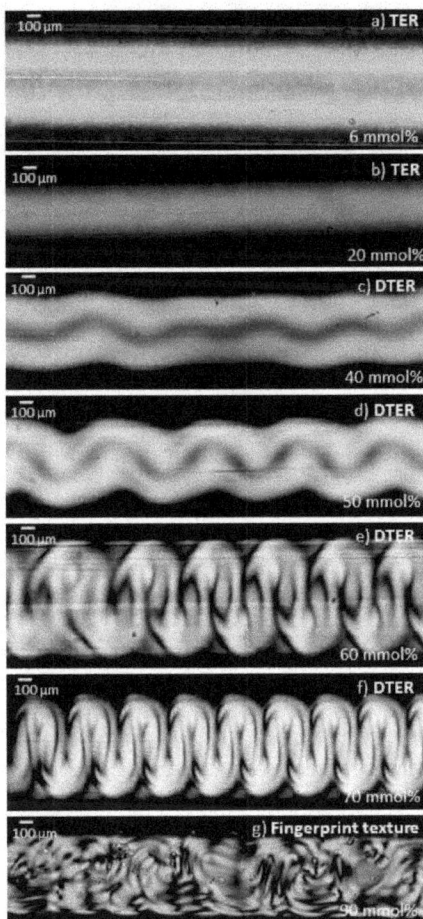

Figure 3.22: Overview of the development of the cylindrical confined LLC structure as a function of the (R)-mandelic acid concentration under the absence of a magnetic field. (a) and (b) show a homochiral (left-handed) TER configuration in the concentration range of ≈ 20 – 30 mmol%, the corresponding director field is given in Figure 3.23 (a). The color change is due to the addition of the chiral dopant, changing the birefringence of the material. (c) – (f) Left-handed doubly twisted escaped radial (DTER) configuration appearing in the concentration range from 40 to 80 mmol% of (R)-mandelic acid. By increasing the chirality, the amplitude and the frequency of the sinusoidal structure are increased as well. (g) Increasing the dopant concentration to 90 mmol% and further, a discontinuous transition to the well-known fingerprint texture is observed.

Figure 3.23 illustrates a possible scenario of the transformation from a left-handed (LH) twisted escaped radial (TER), shown in Figure 3.23a, to a left-handed doubly twisted escaped radial (DTER) configuration by adding a helical twist of the escape direction. The escape direction is represented by blue hats where the director field is in the surface pointing towards the top of the hat. In the case of the TER, the diameter of the capillary corresponds to the twist axis, but as the dopant concentration increases and the pitch of the chiral nematic gets of the order of the diameter, the TER becomes unstable and the escape direction spirals around the capillary axis. The resulting DTER configuration resembles a phone cord. For even higher dopant concentrations, the DTER becomes unstable and transforms discontinuously into the well-known fingerprint texture.

In literature the spiraling twisted escaped radial configuration was already reported once by Kitzerow et al. in the case of a thermotropic LC confined to a cylindrical cavity with homeotropic boundary conditions and doped with a chiral dopant having a temperature-induced twist inversion.[94] To demonstrate the similarity, polarized optical micrographs reprinted with permission from ref. [94] are shown in Figure 3.25a. They used capillaries with diameters of 25, 50, 75 and 100 µm and studied the evolution of the director configuration as a function of the capillary radius R over pitch P. At $R/P \approx 0.5$ the twisted escaped radial configuration started spiraling around the capillary axis and with increasing dopant concentration the sinusoidal amplitude and frequency increased as well, as it happened in case of the LLC shown in Figure 3.22c–f. At higher concentrations, however, the distortions became non-sinusoidal, but still periodic and very complex.[94]

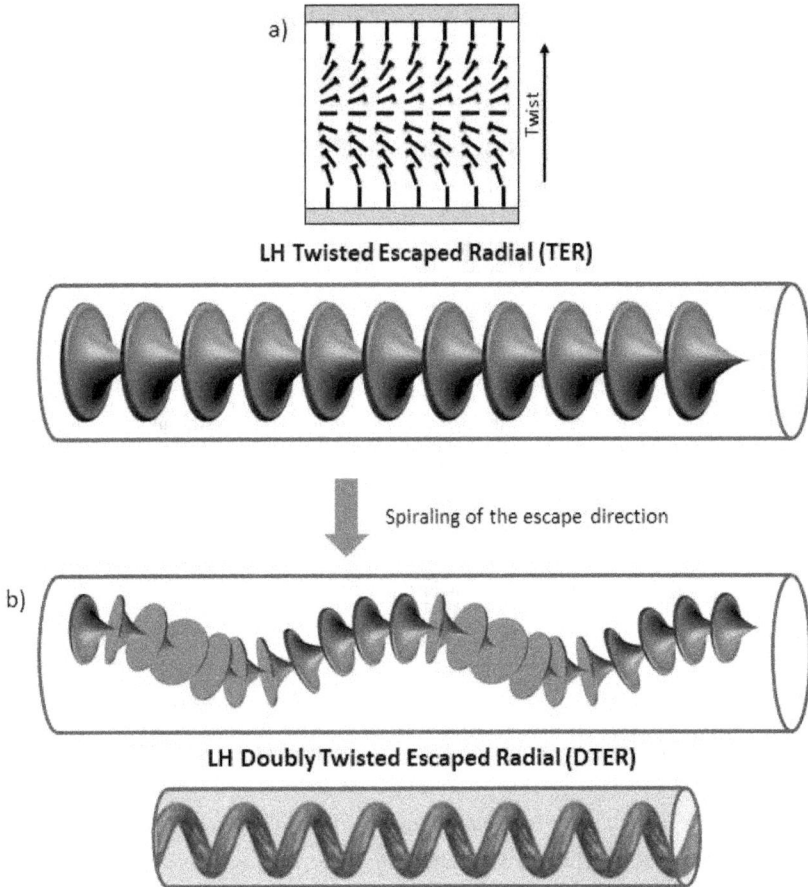

Figure 3.23: Schematic illustration of a TER to DTER transformation scenario. (a) Left-handed (LH) twisted escaped radial (TER) configuration with the twist axis being along the diameter of the capillary. Below: Schematic representation of the escape direction by blue hats. (b) By a spiraling of the escape direction along the capillary axis, a left-handed (LH) doubly twisted escaped radial (DTER) configuration is formed. Below: The spiral of the escape direction resembles a phone cord (red line).

3.4.2.2 Regimes of chiral induction for the Twisted Polar Configuration (H > H_c)

In the following, the effect of a chiral dopant on the twisted polar configuration (TP) is discussed. The capillaries filled with LLC doped with (R)-mandelic acid (same concentrations as used in the previous chapter) were put into a magnetic field in order to generate the twisted polar (TP) configuration. The TP exhibits a twist of the director along the capillary (axial twist) and along the diameter of the capillary (transverse twist), see section 3.4.1.2. The axial and transverse twist deformations have the same handedness and they are mutually coupled. If the axial periodicity (TP pitch) goes towards infinity, the two disclination lines become parallel with a 180° twist along the diameter, assuming that the ½ disclinations are still twist disclinations and not wedge disclinations. Otherwise, if the TP pitch goes to zero, the transversal twist becomes zero (infinite pitch). In addition to the axial and transverse twist deformations, there are local twist deformations around the ½ disclinations, that is why they are called twist disclinations. They have the opposite handedness compared to the axial and transverse twist sense.

The evolution of the TP configuration under the addition of a chiral dopant depends on the characteristic lengths in the system, i.e. the capillary diameter $\emptyset$, the pitch of the TP double helix p_{TP}, the intrinsic pitch of the bulk chiral nematic LLC p, and the distance l between the two disclination lines. According to literature, (R)-mandelic acid has an *HTP* of -330 mm^{-1} for the lyotropic N_D host phase of CDEAB/DOH/H$_2$O,[37,92] meaning that (R)-mandelic acid induces a left-handed director twist, therefore the inverse of the absolute pitch value $|P|^{-1}$ was graphed versus the molar fraction of (R)-mandelic acid x_{MA} in Figure 3.24. Note that in ref. [30,37,92,93] the molar fraction of the mandelic acid does not take the water of the system into account, whereas the molar fractions given here are with respect to all components of the system, including the water content. In the concentration range of 6 mmol% - 200 mmol% of mandelic acid different regimes of chiral induction were observed, as illustrated in Figure 3.24.

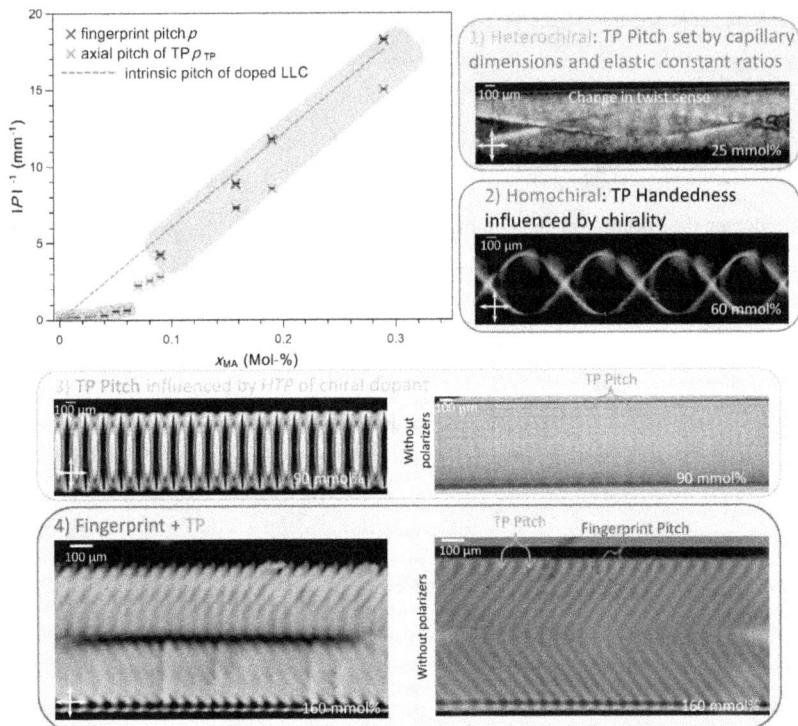

Figure 3.24: Overview of the structural evolution of the LLC twisted polar configuration (TP) in $\emptyset = 700$ μm capillaries as a function of the (R)-mandelic acid concentration. In the diagram, the inverse of the absolute pitch value IPI^{-1} is graphed versus the molar fraction of (R)-mandelic acid x_{MA}. The error is given by the standard deviation of the mean. The IPI^{-1} values originating from the axial TP pitch p_{TP} are marked in green, whereas those originating from the fingerprint pitch p are marked in red. The linear increase of the inverse of the intrinsic pitch of the doped bulk LLC according to the *HTP* of the chiral dopant is marked as a dashed orange line. The actual evolution of the helix under capillary confinement can be divided into four regimes. 1) Heterochiral regime, $p \gg \emptyset$ left- and right-handed TP domains occur with the same probability. 2) Homochiral regime, the sign of the *HTP* governs the handedness of the TP, but still $p \neq p_{TP}$. 3) TP pitch is significantly influenced by the *HTP*, $p \approx p_{TP}$. The image in Figure 3.24 without polarizers confirms the existence of the two ½ disclination lines forming a double helix (TP pitch). 4) TP configuration becomes unstable and a fingerprint texture develops along the capillary axis with a fingerprint pitch p, p_{TP} cannot follow p anymore because of $p < p_{TP}$. The image without polarizers confirms the existence of the TP double helix formed by the two ½ disclination lines and the fingerprint pattern.

At zero dopant concentration, the TP pitch is governed solely by the capillary diameter and the elastic constant ratios. The two $+ \frac{1}{2}$ disclination lines form a double helix because a wedge disclination costs more energy than a twist disclination due to that the twist elastic constant of the LLC is small compared to the splay and bend elastic constants. Because the system is still overall achiral, left- and right-handed TP sections occur with the same probability.

When doping the system with chiral dopant molecules, there is an intrinsic tendency to form a helix with a pitch p normal to the director. This intrinsic pitch p is related to the concentration and the *HTP* of the chiral dopant. In the regime of very low dopant concentrations where $p \gg \varnothing$ the axial periodicity of the TP double helix p_{TP} is thus still mainly governed by $\varnothing$ and the elastic constant ratios. Right- and left-handed TP domains still coexist and no increase of left-handed TP domains at the expense of right-handed TP domains could be observed. Hence, in the concentration range from 0 - 30 mmol% the system is still heterochiral. However, the p_{TP} values might be very weakly affected by the *HTP* of the chiral dopant, but these variations are within the range of expectations.

Increasing the dopant concentration further, p decreases and the influence of the *HTP* on the p_{TP} becomes significant. Above a certain threshold of dopant concentration, it becomes too costly for the system to form a TP double helix with the opposite axial twist handedness compared to the one favored by the dopant. At this point, the system becomes homochiral. It should be pointed out that it is the handedness given by the sign of the *HTP* that governs the handedness of the homochiral TP helix. However, the intrinsic pitch p still does not match p_{TP}. The local twist deformation around the disclination lines is still irrelevant for the sign of p_{TP}. With further increase of the dopant concentration, sooner or later a regime is entered where $p \approx p_{TP}$. At this crossover point, the axial twist increases while the transverse twist decreases – meaning that the transverse pitch increases while the axial pitch decreases. When the transverse pitch becomes larger than the intrinsic pitch p, the transverse twist starts to counteract the dopant-induced contraction of the axial pitch. In order to minimize the elastic penalty related to the transverse twist, the disclination lines are pushed towards the capillary walls, giving an increase in the distance l between the two $+ \frac{1}{2}$ disclination lines approaching the diameter $\varnothing$ of the capillary.

At higher dopant concentrations (≥ 0.1 mol%), p_{TP} cannot any longer follow p because of the increasing elastic penalty from the (unwinding) transverse twist. Furthermore, there will be a rapid increase in elastic energy due to the fact, that the distance between the disclination lines and the capillary walls becomes very small. Therefore, at some point, p_{TP} does not respond anymore to a further increase of dopant concentration, while p continues to decrease. When $p <$ p_{TP}, the TP configuration becomes unstable and a fingerprint texture develops along the capillary axis, with a helical pitch p. At the capillary surface, however, the helix period is still set by the now fixed p_{TP}. Hence, in this concentration regime, the situation is that the bulk cholesteric pitch p decreases without the possibility for the surface p_{TP} to follow. This leads to a kink in the fingerprint structure, i.e. the cholesteric pitch tilts away from the capillary long axis and a "conical chevron" in the fingerprint texture at the capillary center is obtained.

In fact, the instability and formation of the chevron fingerprint configuration can be understood with an analogy to the well-known vertical chevron formation in smectic C bookshelf cells.[21,95–101] In the so-called bookshelf geometry, the smectic layers are vertically arranged between the two glass plates like books in a bookshelf. When the smectic A to smectic C transition occurs in such cells, the layers kink and form a chevron. The reason is that the tilting of the director at the transition evokes that the smectic periodicity shrinks, while the smectic A periodicity stays fixed at the surface due to surface pinning.[21,95–101]

Transferring this to the tilted fingerprint under capillary confinement, the smectic periodicity corresponds to the cholesteric pitch p while the pinned smectic A periodicity at the surface corresponds to the locked p_{TP}. Figure 3.25b shows the example of a field-induced horizontal chevron structure of the SmC* helical structure of a thermotropic epoxide LC in a bookshelf cell.[99] The disclination lines due to the SmC* helix are parallel to the smectic layers. In the region where two domains of opposite layer tilt meet, a vertical chevron defect forms the domain boundary.[99] This texture in Figure 3.25b looks very similar to the texture shown in Figure 3.24, figure section 4).

There can be several explanations discussed why the marginal p_{TP} near the inner capillary surface cannot decrease further at a certain dopant concentration while the cholesteric pitch p in the bulk can. First, as mentioned above, the increases of elastic energy from the unwinding transverse twist and due to the shrinking distance between disclination lines and the inner capillary surface can prevent a further decrease of p_{TP}. Then, the number of ions (of the surfactant) could be lower at the capillary wall than in the bulk, because their solvation sphere would be truncated by the glass surface. This would lead to a disturbance of the very sensitive interplay between the host phase and the chiral dopant molecules.

In addition, the mandelic acid molecules might assemble themselves into dimers (like lots of carboxylic acids do) and become comparably nonpolar which means that the polar glass surface would be avoided by these species leading to a decrease in the mandelic acid concentration at the inner capillary glass surface. Additionally, according to the Gibbs adsorption isotherm for multicomponent systems the concentration ratios have to change at the interface.[102] In both cases, the p_{TP} would experience a different situation regarding the chiral induction process as the cholesteric pitch p in the bulk would do.

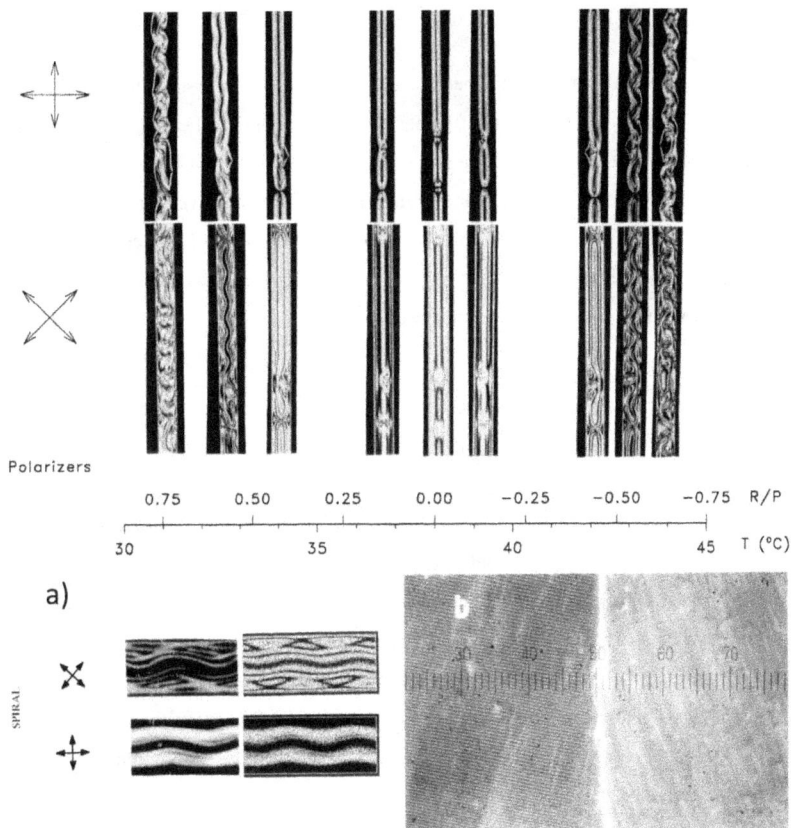

Figure 3.25: (a) Top: Evolution of the director configuration of a thermotropic cholesteric LC in a capillary as a function of the capillary radius R over pitch P. The *Helical Twisting Power* of the chiral dopant has a temperature-induced sign inversion of the pitch. The photographs are positioned over the corresponding bulk R/P ratios. The arrows on the left indicate the arrangement of crossed polarizers for each row. P^{-1} is linearly related to the temperature as shown on the temperature scale. (a) Bottom: Observation and calculated director pattern of the doubly twisted escaped radial configuration (DTER). Left column: Microscope photograph with the arrangement of the polarizers indicated by the arrows. Right column: model transmission pattern. The pictures of (a) are reprinted with permission from ref. [94]. (b) Texture micrograph of a field-induced horizontal chevron structure of the SmC* helical structure of a thermotropic epoxide LC in a bookshelf cell. The disclination lines due to the SmC* helix are parallel to the smectic layers. In the region where two domains of opposite layer tilt meet, a vertical chevron defect forms the domain boundary. The micrograph in (b) is reprinted with permission from ref. [99].

3.5 Chapter Conclusion

In conclusion, this chapter gives to the best of my knowledge a first example of reflection symmetry breaking under capillary confinement in micellar lyotropic liquid crystals. Confining the director field of the N_D phase of the ternary system of CDEAB/DOH/H$_2$O to a cylindrical geometry, two reflection symmetry broken configurations were observed – namely the twisted escaped radial (TER) and the twisted polar (TP) configuration. The defect-free twisted escaped radial (TER) was observed without the action of a magnetic field, whereas the twisted polar (TP) configuration containing two twisting ½ disclination lines was formed under the action of a magnetic field.

Those configurations are very similar to what has been recently found for chromonic LLCs,[61] which supports the idea that the breaking of reflection symmetry is a general phenomenon in LLCs. The occurrence of reflection symmetry broken configurations of the chromonic N_C phase under capillary confinement was explained by an anomalously low twist elastic modulus – which is one order of magnitude smaller compared to the splay and bend moduli.[56,67,68] This means, that energetically costly splay and/or bend deformations can be avoided by escaping into more favorable twisting deformations instead, leading to helical structures although the system is achiral.[61] The fact that similar reflection symmetry broken configurations were observed for a N_D phase of a non-chiral classical micellar LLC suggests the idea that micellar nematic LLCs should have the same anomaly of elastic constants as chromonic nematic LLCs. This leads to the suspicion that a low twist elastic modulus is a general phenomenon in lyotropic nematic liquid crystals – in contrast to thermotropic nematic LCs, for which no chiral configurations have been reported in similar studies using such as *n*-(4-methoxybenzylidene)-4-butylaniline (MBBA) in the same type of capillaries with homeotropic boundary conditions.[103] In the following chapter, the elastic constants for the N_D phase of the micellar LLC CDEAB/DOH/H$_2$O are measured via depolarized dynamic light scattering, as it was performed by Zhou et al.[67].

This small twist elastic constant, in combination with the geometry of capillary confinement – which permits a formation and propagation of a helix axis along the long axis of the capillary, where helix periods in the centi-/millimeter range can be developed unhindered – gives rise to an extremely high sensitivity for chirality. A concentration of (*R*)-mandelic acid of 30 mmol% is sufficient to bias the twist sense of the twisted polar configuration throughout the centimeter length scale of the whole capillary. Roughly speaking, just one out of 3000 molecules is chiral in this case. This phenomenon can be related to the so-called "sergeants-and-soldiers" effect, which was first proposed in the field of polymer chemistry. [104,105] In a system with "sergeants-and-soldiers" behavior a chiral additive (the "sergeant") imposes its chirality on a structure formed by achiral components (the "soldiers").[106-108]

4 Viscoelastic properties of micellar lyotropic liquid crystals

Chapter Overview

In the previous chapter, the spontaneous formation of chiral structures in a lyotropic liquid crystal (LLC) under capillary confinement was reported.[45] Due to the fact, that the system consists only of achiral components, this is an example of spontaneous reflection symmetry breaking. Recently, similar chiral configurations were found for the special case of chromonic LLCs and their occurrence was attributed to an anomalously low twist elastic modulus, which was measured to be one order of magnitude smaller than the other two moduli in the nematic phase.[61,67,68] This is in contrast to what is known for thermotropic LCs in which all elastic constants are in the same range.[68] Comparing the results from the chromonic LLCs with the ones of our standard micellar LLC, the experimental observations suggest that micellar LLCs might have the same anomaly of elastic constants as was found for chromonic LLCs. In this chapter, a light scattering study gives experimental evidence that the twist elastic constant of the micellar nematic phase is indeed surprisingly small. Following the procedure of Zhou et al.[68], the measurements were performed with a depolarized light scattering setup in which light is scattered by nematic director fluctuations, the magnitude of which depends on the three elastic constants as well as on the viscosities of the nematic phase.

4.1 Elastic constants

4.1.1 Measuring elastic constants

The three elastic constants splay K_1, twist K_2, and bend K_3, play an important role in all types of bulk deformations of the nematic director field. In practice, in order to measure elastic constants, director deformations are either generated by external magnetic or electric fields or imposed by surface anchoring conditions in nontrivial confinements.[109] The elastic response can be probed optically, by capacitance measurements, by thermal or electric conductivity measurements or with nuclear magnetic resonance (NMR).

The standard method for measuring elastic constants in nematics is based on the Fréedericksz transition analyzing one-dimensional director deformations in thin cells with defined anchoring conditions.[110–112] The initially homogenous director field in the sample is distorted by an external electric or magnetic field only above a certain threshold field strength, the so-called Fréedericksz threshold, which depends upon the elastic properties of the liquid crystal. However, there are more elegant ways to measure elastic constants without inducing static or dynamic director deformations by applying external forces and where only one type of surface anchoring is needed. For example, studying simply the thermal fluctuations by means of scattering techniques or NMR spectroscopy can also give information on the elastic moduli. Light scattering of nematics is due to thermal orientational fluctuations of the director field which depend on elastic constants and viscosities of the material.[109,113]

4.1.2 The elastic constants of lyotropic and thermotropic LCs

In the literature, the elastic constants of thermotropic liquid crystals are well known.[34,82,109,113,114] On the contrary, the elastic constants of lyotropic liquid crystals have not been studied extensively so far.[67,68,115–118] Given the difficulties when applying an electric field to LLCs, this complicates the common measurements via the Fréedericksz transition. However, Saupe et al.[118] confined the lyotropic N_D liquid crystal CsPFO/H$_2$O to a cylindrical capillary controlling the structure of the sample by means of a magnetic field and determined the elastic constants K_1 and K_3 and the rotational viscosity of the lyotropic N_D liquid crystal CsPFO/H$_2$O by conductivity measurements. A small ac- voltage was applied along the long capillary axis and the conductivity constants were obtained from static measurements at different magnetic fields.[118] The values of K_1 and K_3 from this publication are listed in Table 4.1.

In 2012, Zhou et al. used a magnetic Fréedericksz transition technique to measure the temperature and concentration dependences of the splay, twist and bend elastic constants for the nematic phase of the chromonic LLC sunset yellow.[67] For this technique, the diamagnetic susceptibilities are required and the LC director has to be uniformly aligned in both a planar and a homeotropic way. In homeotropically aligned cells, the bend constant can be measured, whereas planar aligned cells are needed for the splay and twist constants. Since either homeotropic or planar alignment is difficult to realize in lyotropic systems depending on the building block shape, this magnetic Fréedericksz transition technique is experimentally rather challenging.

In 2014, Zhou et al.[68,115] developed another method to determine the elastic constants of chromonic LLCs which requires only planar surface anchoring in the LC cells. This technique is based on the dynamic light scattering (DLS) approach used by Meyer et al.[116] to characterize the viscoelastic properties of a nematic phase of polymeric LLCs and explained in detail in the following experimental methods section. Meyer et al. studied the lyotropic polymer PBG (Poly-γ-benzyl-glutamate).[116] They used the two geometries shown in Figure 4.2 and measured ratios of elastic constants and viscosities.

However, to get the absolute values, an independent measurement of one of the material parameters is needed, such as K_3 determined by the Fréedericksz Transition. In combination with the diamagnetic anisotropy values for PBG, the three elastic constants, as well as the viscosities, could be deduced. Zhou advanced this technique by calibrating the set-up with the dynamic light scattering measurement of the well-known thermotropic 5CB material such that no independent measurement of one of the material parameters is needed. The known elastic constants of LLCs are listed in Table 4.1.

It has to be pointed out, that because scattered light is detected outside the sample, the birefringence of the nematic has to be taken into account. Meyer et al. stated that if the refractive indices are unknown, the light scattering technique is only applicable in the limiting case of small optical-frequency dielectric anisotropy, and therefore small optical birefringence.[116] The impact of a large birefringence requires the precise measurements of the ordinary and the extraordinary refractive indices, whereas for low birefringent materials ($\Delta n \approx 0.001$) multiple scattering effects can be neglected and yet the intensity of the scattered light is strong enough for detection. Meyer et al. even stated that corrections to the scattering wave vector due to finite optical birefringence remain less than 2% for total scattering angles > 2°.[116]

Table 4.1: Overview of what is known from the literature for splay (K_1), twist (K_2) and bend (K_3) elastic constant values for the thermotropic 5CB, and in comparison to that, for lyotropic systems. If not stated otherwise, measurements were performed at room temperature.

Nematic LCs	K_1 /pN	K_2 /pN	K_3 /pN
Thermotropic 5CB[119]	6	3	8
Lyotropic micellar CsPFO/H$_2$O at 309 K[118]	≈ 2	-	≈ 8
Lyotropic polymeric PBG[116,120]	4.1	0.36	4.7
Lyotropic chromonic disodium cromoglycate[68,115]	10	0.7	24
Lyotropic chromonic sunset yellow[67,68]	4.3	0.7	6.1
Lyotropic micellar CEDAB/DOH/H$_2$O	unknown	unknown	unknown

As Table 4.1 shows, the values of all three elastic constants of thermotropic 5CB are in the same order of magnitude. Whereas in the lyotropic systems investigated to date – namely the polymer and the chromonic LLCs – the twist elastic constant K_2 is remarkably one order of magnitude smaller than the other two elastic moduli. A possible reason for the smallness of K_2 is still to be found. However, this result explains the occurrence of reflection symmetry broken configurations of the chromonic LLC sunset yellow under capillary confinement, replacing energetically costly splay and bend deformations with twist deformation by forming equilibrium helical structures.[61] Comparing the results from the chromonic LLCs with the ones of our standard micellar LLC in the previous chapter, the experimental observations suggest that micellar LLC might have the same anomaly of elastic constants as was found for chromonic LLCs. Therefore, the elastic constants of the micellar LLC CDEAB/DOH/H$_2$O have been measured according to the technique from Zhou et al.[68,115].

4.2 Measuring elastic constants via dynamic light scattering

4.2.1 Experimental Setup

According to the Maxwell equations, the propagation of light is sensitive to fluctuations in the dielectric tensor. In liquid crystals, these fluctuations originate from two sources. On the one hand, fluctuations in the magnitude of the dielectric constants parallel or perpendicular to the director **n**, due to small, local changes in the density or temperature, etc.; or on the other hand, fluctuations in the orientation of **n** can cause fluctuations in the dielectric tensor as well, which is – specific to the nematic phase – the dominant effect. The thermally excited angular fluctuations evoke strongly depolarized light scattering with high angular asymmetry. Therefore, the intensity of the scattered light depends strongly on the geometry of the experiment – the polarization of the incident and scattered light as well as the orientation of the director in the sample – and the angle of detection θ.[113]

A schematic light scattering geometry is shown in Figure 4.1. The scattering plane contains the scattering vector q, the scattering angle θ, the wave vectors k_i and k_s of the incident and scattered light, respectively. There is scattered light parallel and perpendicular to the scattering plane, but the input polarization can be either in the scattering plane or perpendicular to it.[113] In the following experiment it was chosen to be perpendicular.

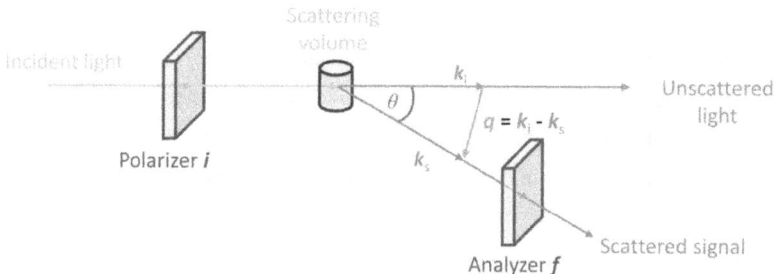

Figure 4.1: Schematic illustration of the geometry of a light scattering experiment. The sketch is redrawn based on ref. [113].

Choosing between a setup in which the input polarization of the electric field vector **E** is either parallel or perpendicular to the scattering plane gives two possible configurations.

If the polarization of the measured scattered signal is perpendicular to the electric field vector **E** of the incident light, this is a so-called depolarized light scattering setup. In the literature, a depolarized setup is also called "vh-geometry" with vh meaning "vertical horizontal".[113] Typically in a "vh" experiment, the electric field vector **E** of the incident laser beam is vertically polarized, whereas only horizontally polarized light is detected. A polarized light scattering setup could be called "vv" meaning "vertical vertical".[113] Having a nematic sample with a homogenous planar alignment of the director **n**, and choosing the "vh-geometry", there are two different ways the director **n** can be aligned with respect to the polarization of the **E** vector of the incident light. When the director is parallel to the input polarization, splay and twist deformations can be observed (splay-twist geometry). Whereas when the director is perpendicular to the input polarization, bend and twist deformations can be detected (twist-bend geometry), for more information see appendix.[14,113,121] A schematic illustration of the splay/twist and bend/twist geometries is shown in Figure 4.2.

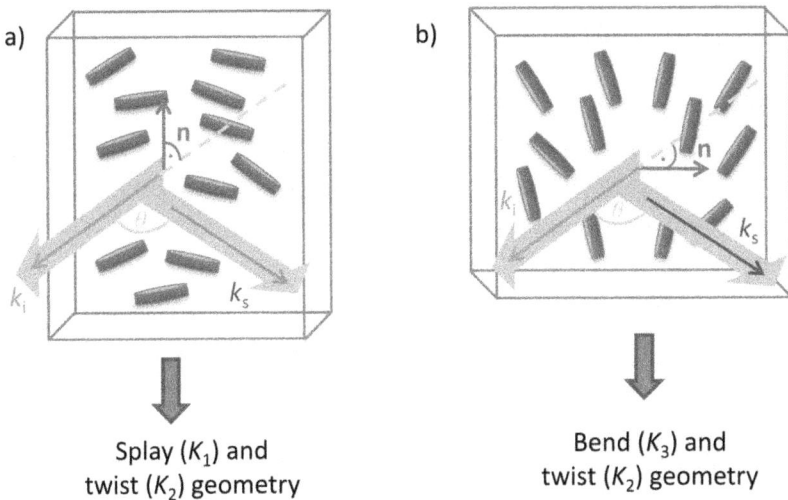

Splay (K_1) and twist (K_2) geometry

Bend (K_3) and twist (K_2) geometry

Figure 4.2: Schematic illustration of the two director geometries in the depolarized ("vh") light scattering setup. (a) Splay-twist geometry (pure splay is shown) and (b) twist-bend geometry (pure bend is shown). In (a) the nematic director **n** is aligned perpendicular to the polarization of the measured scattered light, whereas in (b) **n** is parallel to the scattering plane. The sketch is redrawn based on ref. [68].

4.2.2 Sample preparation

The unidirectional planar alignment of the nematic sample is achieved by planar anchoring conditions at the inner glass surfaces of cells into which the nematic LC is filled. These cells were produced in the MC2 cleanroom facility of the Chalmers University of Technology in Gothenburg. The ITO coated glass plates were first cleaned in a megasonic bath of NH_3, H_2O_2, and H_2O at 100 °C for 10 minutes, four times each rotating the plate every round by 90°. After rinsing with H_2O and drying with N_2, the plates were spin-coated with polyimide PI-2610 (DuPont). Afterwards, the plates were soft baked by placing them on a hot plate at 110 °C for 1 minute, followed by a curing process in which the plates were dried for 3 hours in an oven at 300 °C. Then they were subsequently rubbed with a velvet cloth in a commercial rubbing machine (LC-Tec Automation); see Figure 4.3b. Finally, the substrates were glued together by means of UV-curing glue with 30 µm and 100 µm diameter spherical silica spacers defining the cell thickness and assembled in an aligning and assembling machine (CIPOSA), see Figure 4.3a. 6 mm x 6 mm sized cells were obtained by scribing and breaking. The 30 µm cells were used for 5CB in order to minimize possible multiple scattering, whereas due to the low birefringence of the LLC 100 µm cells were used for the lyotropic samples.

The cells are filled with liquid crystal by capillary forces at room temperature, and the edges are sealed with UV glue (Norland Optical Adhesive 71) to prevent solvent evaporation. In order to eliminate possible alignment effects from the filling process, the samples are heated into the isotropic phase and cooled down to the nematic phase. In the center of Figure 4.4a, an example of a homogenously planar aligned LLC is shown. Due to the relatively large cell thickness, the alignment can take 1 – 2 days before it is uniform. A sample holder for the cells to put into the index matching fluid bath of the light scattering setup was manufactured by the workshop of the Institute of Physical Chemistry; see Figure 4.4b. In the DLS setup, the perpendicular alignment of the incoming laser beam with the cell was verified by the reflection of the incident laser beam at the ITO coating. An example of a cell filled with 5CB is shown in Figure 4.4c. The small dimensions of the cell complicate the filling and sealing process.

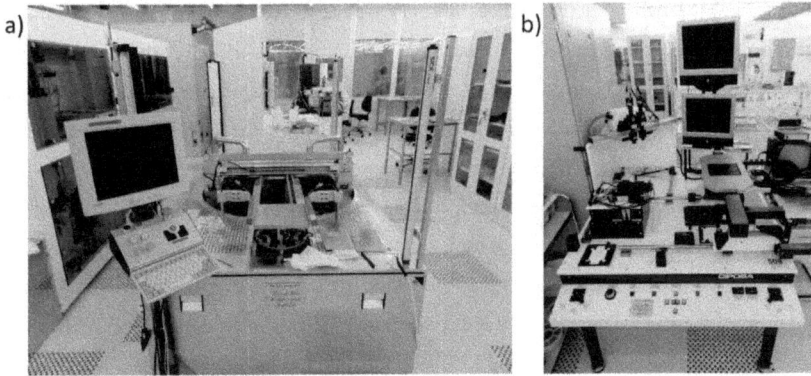

Figure 4.3: Experimental equipment for producing LC cells in the MC2 cleanroom of the Chalmers University of Technology in Gothenburg. (a) The machine used for rubbing the polyimide coated glass substrates with a velvet cloth. (b) Aligning and assembling machines for placing two glass substrates together and fix them by UV-curing of the adhesive.

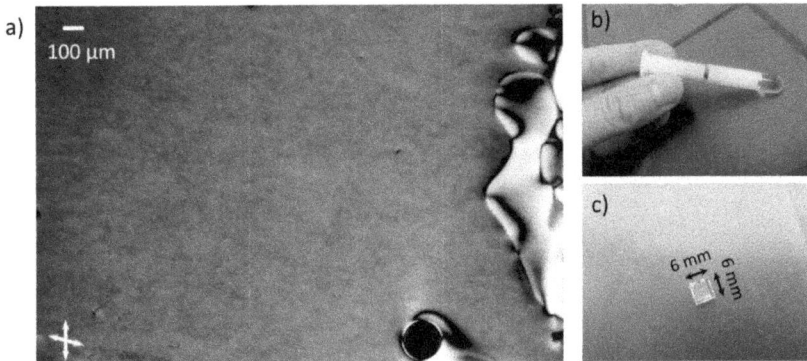

Figure 4.4: (a) Planar aligned LLC in a 100 µm thick cell. The center of the cell is without defects and therefore suited for dynamic light scattering measurements. (b) Sample holder for the cells manufactured by the workshop of the Institute of Physical Chemistry to put into the index matching fluid bath of the LS setup. (c) An example of a cell filled with 5CB and sealed with UV glue is shown.

Furthermore, the two refractive indices $n_\parallel$ and $n_\perp$ have to be measured. This was done by interferometry according to ref. [122]. The basic measurement concept is the occurrence of reflection at both inner glass surfaces (glass thickness > 50 µm) which are separated by a small distance ($d = 2 - 25$ µm). This leads to an interference pattern in the visible light, which can be detected spectroscopically in transmission. If the sample is irradiated with polarised light which is parallel or perpendicular to the director of the homogenously planar aligned LC phase, the two refractive indices can be obtained for a known cell thickness.[122] For measurements, nylon rubbed LC cells with a thickness of 10 µm were used to achieve a unidirectional planar alignment of the LLC. The refractive index $n_\parallel$ of the LLC was measured to be 1.396±0.0001, whereas the refractive index $n_\perp$ of the LLC was measured to be 1.392±0.0001. The resulting birefringence of $\Delta n \approx 0.004$ was confirmed by measuring the optical retardation with a LC-PolScope equipment.

4.2.3 Dynamic light scattering (DLS) analysis

4.2.3.1 General DLS analysis

The samples were measured with a 2D pseudo cross-correlation dynamic light scattering setup from LS instruments with a high-performance DPL Laser (Cobolt) of $\lambda = 561$ nm wavelength and a maximum power of 400 mW. The polarization of the incident laser light is vertical to the scattering plane, whereas the scattered light passing through a horizontal analyzer was detected. In order to minimize the amount of dust, the lyotropic mixtures were prepared using filtered double distilled water and filtered DOH (pore size of filter = 0.2 μm). Because the same signal is the source for both the autocorrelation as well as the pseudo cross-correlation detector, the term "auto" will be used in the following.

In the subsequent introduction of light scattering data analysis, particles are used as scattering species since they produce fluctuations just as in the case of liquid crystals.[123,124] If the scattering species are moving, fluctuations in the scattered intensity with time reflect the so-called Brownian particle motion of the scattering species which is caused by thermal density fluctuations of the solvent. These fluctuations in the scattered light intensity are due to a change in the interference pattern caused by fluctuating interparticle distances, resulting in a change in the detected scattered intensity with time $I(t)$ measured at a given scattering angle θ. To quantitatively analyze the particle motion by light scattering, the scattering intensity fluctuations are expressed in terms of correlation functions, where the intensity at the beginning is correlated with the intensity at a time t.[123,124]

The intensity $I(t)$ fluctuates in time randomly around a mean value $\langle I \rangle$ with an amplitude that depends on frequency, see Figure 4.5a. Averaging over the measurement time t_m, which is long compared to the fluctuation time τ, gives the value for the mean intensity $\langle I \rangle$. Theoretically, averaging should be done over an infinite time interval, which is not possible due to practical reasons. In dynamic light scattering, the scattering intensity is detected time-dependently as $I(t)$ in very small time intervals Δt ($= \tau/n$ with $0 \leq n \leq N$ and $N\Delta t = t_m$) and used to calculate the autocorrelation function. The time-dependent scattered intensity measured by a detector is multiplied by itself after it has been shifted by a time period τ, and these products are averaged over the total measurement time t_m. Comparing a system of particles at t and $t + \tau$, the difference between the particles with respect to their position, velocity and direction of motion is little for short delay times τ, meaning that $I(t)$ and $I(t+ \tau)$ are highly correlated. For longer delay times the correlation of the intensities is lost. If Δt is sufficiently small, the intensity autocorrelation function can be written as:[123,124]

$$\langle I(t) \cdot I(t+\tau) \rangle = \lim_{t_m \to \infty} \frac{1}{t_m} \int_{t=0}^{t_m} I(t) \cdot I(t+\tau) \; dt \, . \tag{18}$$

In Figure 4.5b a typical exponential decay of the autocorrelation function $\langle I(t)\,I(t+\tau) \rangle$ is shown. The starting value of the correlation function is $\langle I(t)^2 \rangle$ (for $\tau = 0$) and the curve decreases exponentially to the uncorrelated value of $\langle I(t) \rangle^2$. Defining $\langle\,I\,\rangle = 1$, the intensity correlation function should decay from 2 to 1 for scattering particles in solution moving only according to Brownian motion. Therefore Equation (18) is divided by $\langle I(t) \rangle^2$ and the normalized intensity autocorrelation function is obtained:[123,124]

$$g_2(\tau) = \frac{\langle I(t) \cdot I(t+\tau) \rangle}{\langle I(t) \rangle^2} \geq 1 \, . \tag{19}$$

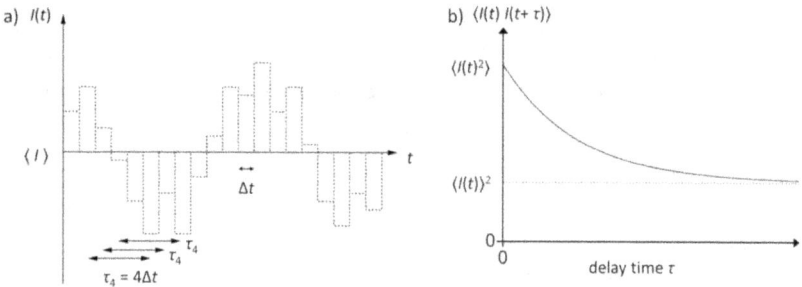

Figure 4.5: (a) In the case of moving scattering particles, fluctuations in the scattered light intensity $I(t)$ occur due to changes in the interparticle interference. $I(t)$ fluctuates time-dependently around a mean value $\langle\,I\,\rangle$ which is calculated by averaging over the total measurement time. The time interval Δt for one measurement has to be much smaller than the fluctuation (delay) times τ. b) shows a typical exponential decay of the autocorrelation function $\langle I(t)\,I(t+\tau) \rangle$, in the case of one relaxation process and hence one decay time. The starting value for $\tau = 0$ corresponds to $\langle I(t)^2 \rangle$ and the curve decreases exponentially to the value of $\langle I(t) \rangle^2$. Sketch is reprinted with permission from ref. [123].

The instrument software gives the value of $g_2 - 1$ which is proportional to the amplitude autocorrelation function g_1 according to the Siegert relation:[123,124]

$$g_2 - 1 = g_1^2 . \tag{20}$$

The amplitude correlation function g_1 is valid for scattered light which exhibits a Gaussian distribution of the electric field amplitude E. For monodisperse systems, g_1 can be expressed by a single exponential function:[123,124]

$$g_1(\tau) = \frac{\langle E(0)\, E(q,\, \tau) \rangle}{\langle E(0)^2 \rangle} = e^{\frac{1}{T}\tau}, \tag{21}$$

with $0 \leq g_1(\tau) \leq 1$, the electric field strength E, the scattering vector q, the decay time τ and the relaxation time T of a single relaxation process. For systems showing more than one relaxation processes with different relaxation times T_i, the amplitude autocorrelation function can be expressed by a sum over weighted exponential functions:[123,124]

$$g_1(\tau) = \sum_{i=0}^{n} B_i\, e^{\frac{1}{T_i}\tau}, \tag{22}$$

with B_i being the weighting factor for the relaxation process with relaxation time T_i.

For relaxation processes showing a distribution of relaxation times, a stretching factor β_i has to be taken into account in the amplitude autocorrelation function:[124,125]

$$g_1(\tau) = \sum_{i=0}^{n} B_i\, e^{-\left(\frac{\tau}{T_i}\right)^{\beta_i}} . \tag{23}$$

The effective relaxation time $T_{eff,i}$ is calculated with the Gamma function $\Gamma(1/\beta_i)$, according to:[125]

$$T_{eff,i} = \frac{T_i}{\beta_i}\, \Gamma\left(\frac{1}{\beta_i}\right) . \tag{24}$$

4.2.3.2 DLS analysis for nematic LCs

4.2.3.2.1 Relative values of the elastic constants

As briefly introduced in Chapter 4.2.1, the scattered light intensity of a nematic LC can be discussed for two different geometries in terms of the normal modes.[14,113] Those two geometries are schematically shown in Figure 4.2. One is the so-called splay-twist geometry, shown in Figure 4.2a, in which the amplitude autocorrelation is a function of fluctuations originating from splay (index 1) and twist deformations (index 2):

$$g_1(\tau) = B_1\, e^{-\left(\frac{\tau}{T_{\mathrm{eff},1}}\right)^{\beta_1}} + B_2\, e^{-\left(\frac{\tau}{T_{\mathrm{eff},2}}\right)^{\beta_2}} . \tag{25}$$

The relaxation times $T_{\mathrm{eff},1}$ and $T_{\mathrm{eff},2}$ are linked to the respective elastic constants K_1 and K_2 and their viscosities η_{splay} and η_{twist} according to:[56,68]

$$\frac{1}{T_{\mathrm{eff},1}(\theta)} = \frac{K_1 q^2(\theta)}{\eta_{\mathrm{splay}}} , \tag{26}$$

$$\frac{1}{T_{\mathrm{eff},2}(\theta)} = \frac{K_2 q^2(\theta)}{\eta_{\mathrm{twist}}} . \tag{27}$$

The scattering vector is expressed as:[124]

$$q(\theta) = \frac{4\pi\, n}{\lambda} \sin\frac{\theta}{2} , \tag{28}$$

as the laser wavelength λ, the scattering angle θ and the mean refractive index n:

$$n = \frac{n_\| + n_\perp}{2} , \tag{29}$$

where $n_\|$ and $n_\perp$ are the two refractive indices for light polarized parallel and perpendicular to the director, respectively.

Equations (25) – (27) demonstrate that a separation of the two modes in the time domain can only be achieved by a substantial difference in either the two viscosities – or the splay and twist elastic constants. On the other hand, if the splay and twist elastic constants and respective viscosities are in the same order of magnitude, the separation cannot be done, as is the case for thermotropic 5CB and other materials which follow the one-constant approximation.

The scattering intensity in the splay-twist geometry I_{12} can be written as follows (for more details see the appendix):

$$I_{12}(\theta) = X \,\Delta n^2 \, d \, P \left[\frac{1}{K_1}\cotan^2\left(\frac{\theta}{2}\right) + \frac{1}{K_2}\right], \tag{30}$$

as a scaling factor X including constant factors like $k_B T$ or the wavelength of the laser λ, Δn as the birefringence of the material, d as sample thickness of the used cell and P as the incident laser power. Under the assumption, that the splay and twist modes can be separated, the splay intensity I_1 and the twist intensity I_2 can be described as:

$$I_1^{\text{lyo}}(\theta) = X_{\text{lyo}} \,\Delta n_{\text{lyo}}^2 \, d_{\text{lyo}} \, P \left[\frac{1}{K_1^{\text{lyo}}}\cotan^2\left(\frac{\theta}{2}\right)\right] \tag{31}$$

and:

$$I_2^{\text{lyo}}(\theta) = X_{\text{lyo}} \,\Delta n_{\text{lyo}}^2 \, d_{\text{lyo}} \, P \left[\frac{1}{K_2^{\text{lyo}}}\right]. \tag{32}$$

Dividing the measured intensity I_i by the incident laser power P is from now on referred to as amplitude A_i.[2]

The fraction of the measured intensity for the splay and twist signal respectively is obtained from the weighting factor B_i of the corresponding relaxation process which is calculated with a Python Script written by Tobias Steinle. Plotting the splay amplitude A_1 versus $\cotan^2(\theta/2)$ gives a line through the origin with the slope $a_{1,\text{lyo}}$:

$$a_{1,\text{lyo}} = X_{\text{lyo}} \,\Delta n_{\text{lyo}}^2 \, d_{\text{lyo}} \, \frac{1}{K_1^{\text{lyo}}}. \tag{33}$$

Plotting the twist amplitude A_2 versus $\sin^2(\theta/2)$ (this is referred to as the q angular dependence in the following) gives a constant value $a_{2,\text{lyo}}$:

$$a_{2,\text{lyo}} = X_{\text{lyo}} \,\Delta n_{\text{lyo}}^2 \, d_{\text{lyo}} \, \frac{1}{K_2^{\text{lyo}}}. \tag{34}$$

---

[2] Actually the amplitude A_i would have the unit $\left[\frac{\text{counts per second}}{\text{mW}}\right]$, but because it cancels out through calibration, it is neglected in the following.

The second geometry is the so-called twist-bend geometry, schematically shown in Figure 4.2b. Expressing the scattering intensity in terms of the normal modes and introducing the angular dependence of the scattering vector q, leads to a term in which the twist and bend modes cannot be separated, for details see the appendix. However, under the assumption that the twist elastic constant K_2 is much smaller than the bend elastic constant K_3, the amplitude correlation function is approximately only dependent on bend fluctuations (index 3). Hence, the correlation function is a single exponential function and for lyotropic systems, this geometry is simply called bend geometry:

$$g_1(\tau) = B_3 \, e^{-\left(\frac{\tau}{T_{\text{eff},3}}\right)^{\beta_3}} . \tag{35}$$

Note that in, e.g. 5CB, this approximation does not hold. The relaxation time $T_{\text{eff},3}$ is linked to the respective elastic constants K_3 and its viscosity η_{bend} according to:[56,68]

$$\frac{1}{T_{\text{eff},3}(\theta)} = \frac{K_3 \, q^2(\theta) \cos^2\left(\frac{\theta}{2}\right)}{\eta_{\text{splay}}} . \tag{36}$$

The scattering intensity in the bend geometry I_3 can be written as follows; for more details see appendix:

$$I_3(\theta) \approx X \, \Delta n^2 \, d \, P \left[\frac{4}{K_3} \cotan^2(\theta)\right] . \tag{37}$$

Plotting the bend amplitude A_3 versus $\cotan^2(\theta)$ gives a line through the origin with the slope $a_{3,\text{lyo}}$:

$$a_{3,\text{lyo}} = X_{\text{lyo}} \, \Delta n_{\text{lyo}}^2 \, d_{\text{lyo}} \, \frac{4}{K_3^{\text{lyo}}} . \tag{38}$$

The ratios of A_2/A_1, A_2/A_3 and A_1/A_3 plotted against different angular functions give the relative values of the elastic constants $K_1 : K_2 : K_3$.

The ratio of A_2 over A_1 is associated with the twist and splay elastic constants according to:

$$\frac{A_2}{A_1} = \frac{K_1}{K_2} \tan^2\left(\frac{\theta_{12}}{2}\right) , \tag{39}$$

with θ_{12} as the scattering vector in the splay-twist geometry.

The ratio of A_2 over A_3 is associated with the twist and bend elastic constants according to:

$$\frac{A_2}{A_3} = \frac{K_3}{4K_2}\tan^2\theta_3 , \tag{40}$$

using θ_3 as the scattering vector in the bend geometry. The ratio of A_1 over A_3 is associated to the splay and bend elastic constants according to:

$$\frac{A_1}{A_3} = \frac{K_3}{4K_1} \frac{\tan^2\theta_3}{\tan^2\left(\frac{\theta_{12}}{2}\right)} . \tag{41}$$

Plotting each amplitude ratios over the respective angular functions give lines through the origin with the slopes $a_{2/1} = K_1/K_2$, $a_{2/3} = K_3/4K_2$ and $a_{1/3} = K_3/4K_1$.

4.2.3.2.2 Absolute values of the elastic constants and viscosities

The analysis in 4.2.3.2.1 gives ratios of the elastic constants of the measured LC. However, to obtain the absolute values of this LC, a calibration with a nematic LC for which the elastic constants are known, e.g. 5CB, has to be performed.[56,68,115] In the splay-twist geometry of a 5CB sample a separation of the two modes cannot be realized since $K_1 \approx K_2$, see Table 4.1. The scattering intensity in the splay-twist geometry I_{12} can be written as in Equation (30). Plotting A_{12} of 5CB versus $\cotan^2(\Theta/2)$ the slope $b_{1,5CB}$ is proportional to the inverse of K_1, whereas the y-intercept $c_{2,5CB}$ is proportional to the inverse of K_2:

$$b_{1,5CB} = X_{5CB} \, \Delta n_{5CB}^2 \, d_{5CB} \, \frac{1}{K_1^{5CB}} , \tag{42}$$

$$c_{2,5CB} = X_{5CB} \, \Delta n_{5CB}^2 \, d_{5CB} \, \frac{1}{K_2^{5CB}} . \tag{43}$$

In the twist-bend geometry of a 5CB sample the twist contribution cannot be neglected, since $K_2 \approx K_3$. Therefore, the scattering intensity in the twist-bend geometry I_{23} for 5CB can be written as follows:

$$I_{23}(\theta) = X_{5CB} \, \Delta n_{5CB}^2 \, d_{5CB} \, P \left[\frac{\cos^2\theta}{K_2^{5CB}\sin^4\left(\frac{\theta}{2}\right) + K_3^{5CB}\sin^2\left(\frac{\theta}{2}\right)\cos^2\left(\frac{\theta}{2}\right)} \right] . \tag{44}$$

Plotting the amplitude A_{23} versus the angular term in the brackets gives a line through origin with a slope $b_{23,5CB}$:

$$b_{23,5CB} = X_{5CB} \, \Delta n_{5CB}^2 \, d_{5CB} . \tag{45}$$

From the A_{12} and A_{23} graphs and Equations (42), (43) and (45) the three X_{5CB} values are extracted and the mean value $\overline{X}_{5CB}$ is calculated. Normalizing the $a_{1,lyo}$, the $a_{2,lyo}$ and the $a_{3,lyo}$ by that mean value $\overline{X}_{5CB}$ gives the absolute values of the three elastic constants of the lyotropic LC:

$$K_1^{lyo} = \frac{\overline{X}_{5CB}\,\Delta n_{lyo}^2\,\dfrac{d_{lyo}}{d_{5CB}}}{a_{1,lyo}}, \tag{46}$$

$$K_2^{lyo} = \frac{\overline{X}_{5CB}\,\Delta n_{lyo}^2\,\dfrac{d_{lyo}}{d_{5CB}}}{a_{2,lyo}}, \tag{47}$$

$$K_3^{lyo} = 4\,\frac{\overline{X}_{5CB}\,\Delta n_{lyo}^2\,\dfrac{d_{lyo}}{d_{5CB}}}{a_{3,lyo}}. \tag{48}$$

Knowing now the elastic constants of the lyotropic LC, the corresponding viscosities can be calculated. From the splay-twist geometry, the viscosities η_{splay} and η_{twist} are obtained from the slopes of the graphs of the inverses of the relaxation times $T_{eff,1}$ and $T_{eff,2}$ versus q^2, see Equations (26) and (27). In the case of the bend geometry $T_{eff,3}$ is plotted versus $q^2\cos^2(\theta/2)$ and the bend viscosity η_{bend} is obtained from the slope of this graph, see Equation (36).

Depolarized dynamic light scattering was measured in steps of one or two degrees ranging from $\theta = 15 - 60°$, in the bend geometry typically from $\theta = 15 - 25°$ and in the splay-twist geometry typically from $\theta = 20 - 60°$. For the lyotropic samples the acquisition time for each value of θ was 2-3 hours. Up to 3-4 different sample cells were measured, sometimes at different locations, depending on how well the cell was sealed. For the 5CB samples the acquisition time for each θ was 15 minutes. The experiments were performed at room temperature.

4.3 Results and Discussion

4.3.1 5CB calibration

In Figure 4.6a-d examples of the amplitude pseudo-cross correlation function $g_I(\tau)$ measured at several scattering angles in the splay-twist (a, b) and the twist-bend (c, d) geometry of 5CB samples are plotted versus the delay time τ.

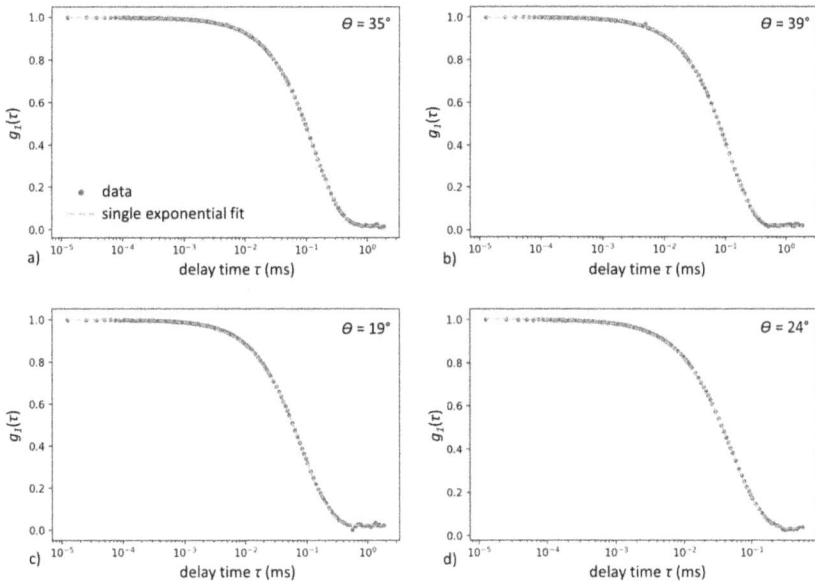

Figure 4.6: (a) – (d) Amplitude correlation function $g_I(\tau)$ plotted versus the delay time τ. (a) Measured in the splay-twist geometry at a scattering angle of $\theta = 35°$ and (b) $\theta = 39°$. (c) Measured in the twist-bend geometry at a scattering angle of $\theta = 19°$ and (d) $\theta = 24°$. The green dots represent the measured data and the yellow dashed line corresponds to the single exponential fit function. Data evaluation and fitting were done with a python script.

Since the splay and twist elastic constant of 5CB are of the same order of magnitude, as well as the respective viscosities, a separation of the splay and the twist mode in the splay-twist geometry cannot be achieved. This means that the amplitude correlation function, as it is shown in Figure 4.6a-b, is fitted to a single exponential function.

In the twist-bend geometry, the amplitude correlation function is fitted to a single exponential function as well; see Figure 4.6c-d. Because they are inherently coupled in the twist-bend geometry, a separation of two modes cannot be achieved. In Figure 4.7 the graphs of the splay-twist amplitude A_{12} and the twist-bend amplitude A_{23} are shown.

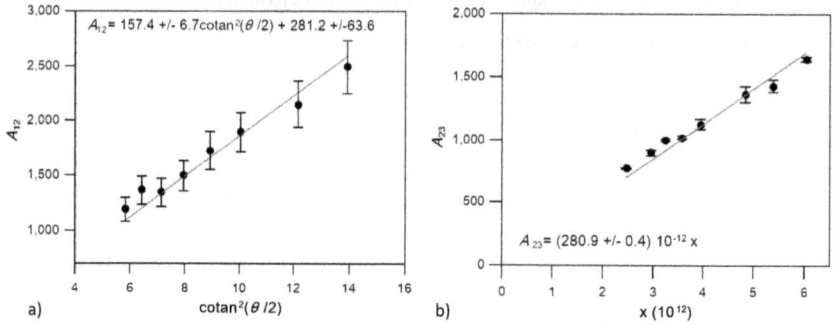

a) $A_{12}= 157.4 +/- 6.7\cot an^2(\theta/2) + 281.2 +/-63.6$

b) $A_{23}= (280.9 +/- 0.4) \, 10^{-12} \, x$

Figure 4.7: (a) splay-twist amplitude A_{12} of 5CB versus $\cot an^2(\theta/2)$. The slope $b_{1,5CB}$ is proportional to the inverse of the splay elastic constant of 5CB, whereas the y-intercept $c_{2,5CB}$ is proportional to the inverse of the twist elastic constant of 5CB. (b) twist-bend amplitude A_{23} of 5CB versus

$$x = \frac{\cos^2\theta}{K_2^{5CB}\sin^4\left(\frac{\theta}{2}\right) + K_3^{5CB}\sin^2\left(\frac{\theta}{2}\right)\cos^2\left(\frac{\theta}{2}\right)}$$ with the slope $b_{23,5CB}$.

From the linear graphs of Figure 4.7, their slopes ($b_{1,5CB}$ and $b_{23,5CB}$) and y-intercept ($c_{2,5CB}$), three X_{5CB} values can be extracted, see Equations (42), (43) and (45). The mean value $\overline{X}_{5CB}$ is calculated to be $(2.9\pm0.2) \, 10^{-8} \, kgs^{-2}$.

4.3.2 K_3 of the lyotropic nematic LC CDEAB/DOH/H₂O

In Figure 4.8a-c the amplitude pseudo cross-correlation function $g_1(\tau)$ measured at several scattering angles in the bend geometry of lyotropic LC samples is plotted against the delay time τ. For further discussion, the phase diagram of the investigated LLC is shown once more in Figure 4.8d.

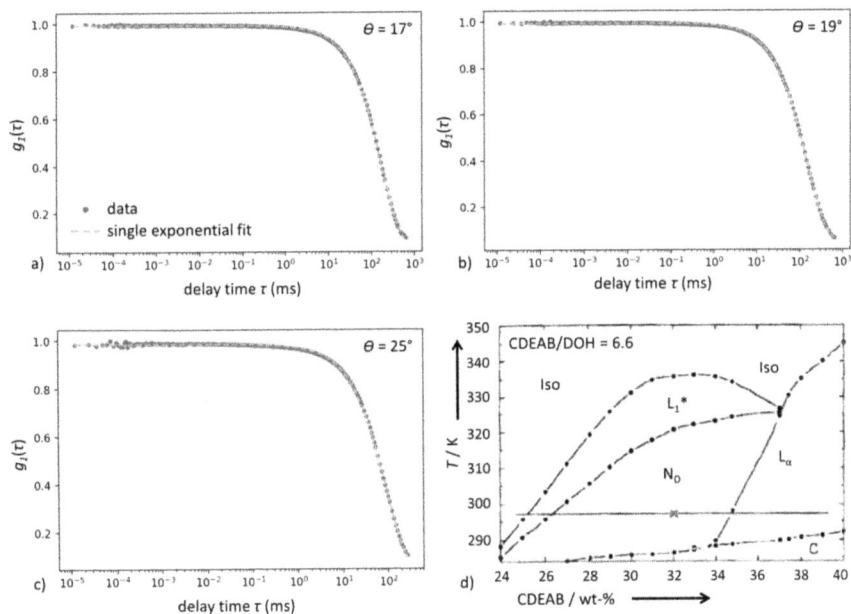

Figure 4.8: (a) – (c) Amplitude correlation function $g_1(\tau)$ at room temperature – measured in the bend geometry at (a) a scattering angle of $\theta = 17°$, (b) $\theta = 19°$ and (c) $\theta = 25°$ – plotted versus the delay time τ. The green dots represent the measured data and the yellow dashed line corresponds to the single exponential fit function. (d) Phase diagram of the ternary system CDEAB/DOH/H₂O for a constant weight ratio of CDEAB/DOH = 6.6. The composition used in this study is marked with a red cross. Symbols: Iso = isotropic phase; C = crystalline phase; N_D = nematic phase with disc-shaped micelles; L_α = lamellar phase; L_1^* = optically isotropic phase with shear-induced birefringence.[80]

As expected, the amplitude correlation function can be fitted by a single exponential fit function, see Equation (35). As discussed above, in lyotropic systems it is only the bend fluctuation mode which can be observed in the (twist)-bend geometry since the twist elastic constant is assumed to be much smaller than the bend elastic constant.

Figure 4.9 shows the linear graph of the bend amplitude A_3 versus the respective angular function. The fitting gives a result of (0.07233 ± 0.00008) kgs^{-2}(pN)$^{-1}$ for the slope $a_{3,lyo}$.

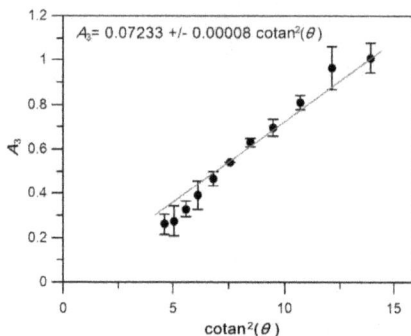

Figure 4.9: Bend amplitude A_3 of the LLC versus cotan$^2(\Theta/2)$. Fitting to a line through the origin gives a slope $a_{3,lyo} = (0.07233\pm0.00008)$ kgs^{-2}(pN)$^{-1}$.

Inserting $\overline{X}_{5CB}$ and $a_{3,lyo}$ into Equation (48) gives a bend elastic constant of (25.7 ± 2.1) pN for the N_D phase of the LLC CDEAB/DOH/H$_2$O. This relatively high value – compared to e.g. K_3 of thermotropic nematic 5CB (8 pN) or polymeric lyotropic nematic PBG (10 pN) – can be explained due to the proximity of the micellar N_D phase of this lyotropic system to a lamellar phase L_α, see phase diagram in Figure 4.8d.

In lyotropic LCs the phase behavior is strongly influenced by the solvent concentration. Therefore, the effect on the phase behavior evoked by a change in composition in lyotropic LCs corresponds to a change in temperature in thermotropic LCs. The investigated CDEAB/DOH/H$_2$O system contains 32 wt% of CDEAB, indicated with a red cross in Figure 4.8d which is only 2 wt% away from the transition to an L_α phase. As discussed by Bajc et al.[118], the bend elastic constant diverges in the vicinity of a nematic-lamellar phase transition, whereas the splay constant shows no critical behavior. This reasoning also applies to the investigated CDEAB/DOH/H$_2$O lyotropic system. It is possible that lamella-like fluctuations exist in the N_D phase, which inhibits bend deformation in the director field.

4.3.3 K_1 and K_2 of the lyotropic nematic LC CDEAB/DOH/H$_2$O

In Figure 4.10 the amplitude pseudo cross-correlation function $g_I(\tau)$ measured at several scattering angles in the splay-twist geometry of the investigated LLC CDEAB/DOH/H$_2$O is plotted against the delay time τ.

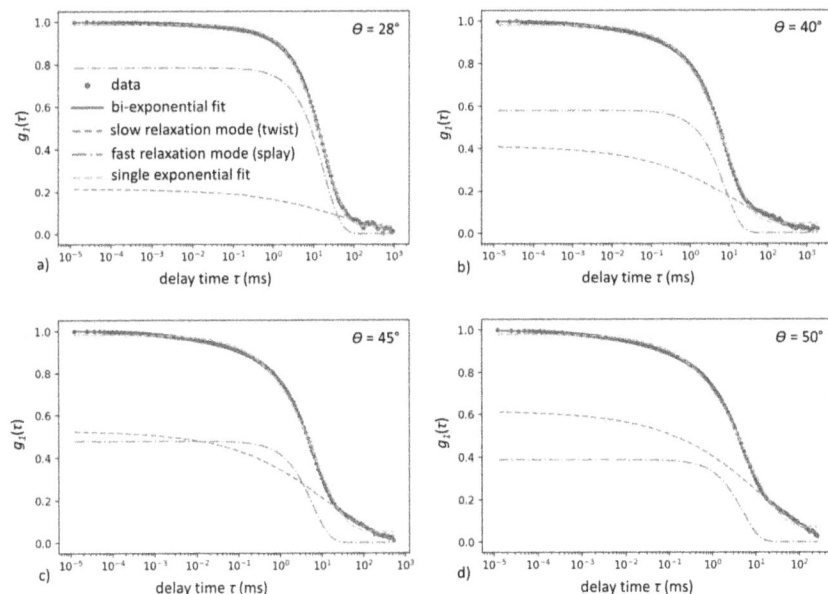

Figure 4.10: (a) – (d) Amplitude autocorrelation function $g_1(\tau)$ at room temperature – measured in the splay-twist geometry at increasing scattering angles (a) $\theta = 28°$, (b) $\theta = 40°$, (c) $\theta = 45°$ and (d) $\theta = 50°$ – plotted against the delay time τ. The green dots represent the measured data and the red line corresponds to the bi-exponential fit function consisting of a slow and a fast relaxation mode (grey dashed lines). The slow relaxation process corresponds to twist fluctuations and the fast relaxation process to splay fluctuations. For comparison, a single exponential fit was tried as well (yellow dashed line).

In contrast to the bend geometry where only one relaxation process was observed, there are now two overdamped relaxation processes in the case of the splay-twist geometry which can be deduced from the kink in the correlation graph. The fact that two relaxation modes can be resolved because of their different delay in time is a first indication that the splay and twist elastic constants are not of the same order of magnitude – contrary to, for example, the case of the thermotropic calibrating substance 5CB where no kink in the correlation function can be observed because splay and twist constants are of the same order of magnitude.

For reasons of comparison, the correlation data in Figure 4.10a-d were fit to a single exponential function, but as one can see, the yellow dashed line misses the data clearly, whereas the red line, corresponding to the bi-exponential fit, describes the data much better, especially the kink in the correlation curves. This confirms the presence of two overdamped relaxation processes. The assignment whether the fast or the slow mode corresponds to either splay or twist fluctuations is done by the angular dependence. According to Equations (30) – (32), the splay mode amplitude is proportional to $\cos^2(\theta/2)$, whereas the twist mode amplitude is proportional to $\sin^2(\theta/2)$. This means, that the fraction of the scattered light intensity should increase for the twist mode with increasing scattering angle θ, whereas it should decrease for the splay mode. Looking at the evolution of the weighting factor for the two relaxation processes in Figure 4.10, the slow mode corresponds to twist fluctuations, whereas the fast mode corresponds to splay fluctuations.[115] Note that around a scattering angle of 45° the fraction of scattered light originating from splay fluctuations is roughly equal the fraction of scattered light originating from twist fluctuations, see Figure 4.10c.

In Figure 4.11 the splay amplitude A_1, the twist amplitude A_2, the addition of those two amplitudes to a splay-twist amplitude A_{12} and the ratio of the twist over splay amplitude A_2/A_1 is graphed versus the corresponding angular functions.

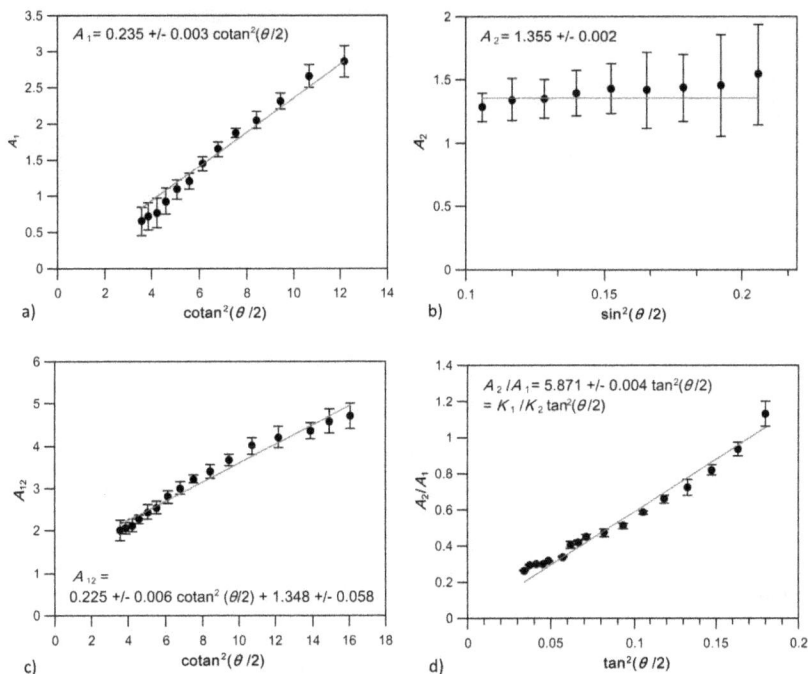

Figure 4.11: (a) Splay amplitude A_1 plotted versus $\cotan^2(\theta/2)$. The fitting gives a slope $a_{1,lyo} = (0.235\pm0.003)$ kgs^{-2}(pN)$^{-1}$. (b) Twist amplitude A_2 plotted versus $\sin^2(\theta/2)$. The fitting gives a constant value of $a_{2,lyo} = (1.355\pm0.002)$ kgs^{-2}(pN)$^{-1}$. (c) Combination of splay and twist amplitude A_{12} plotted versus $\cotan^2(\theta/2)$. (d) Ratio of the twist over the splay amplitude A_2/A_1 graphed versus $\tan^2(\theta/2)$. From the fitting, the K_1/K_2 ratio is estimated to be around 5.871 ± 0.004.

The linear fitting of the data in Figure 4.11a gives a slope of $a_{1,lyo} = (0.235\pm0.003)$ kgs^{-2}(pN)$^{-1}$, whereas the fitting to a constant value in Figure 4.11b gives $a_{2,lyo} = (1.355\pm0.002)$ kgs^{-2}(pN)$^{-1}$. Inserting $\overline{X}_{5CB}$ and $a_{1,lyo}$ and $a_{2,lyo}$ into equation (46) and (47) respectively gives a splay elastic constant of (1.98 ± 0.17) pN and a twist elastic constant of (0.343 ± 0.032) pN for the N_D phase of the LLC CDEAB/DOH/H$_2$O. These values are in agreement with the relative K_1/K_2 ratio obtained from the slope of the linear function shown in Figure 4.11d. To complete the set of ratios of elastic constants among each other, Figure 4.12 shows the graphs of the two other amplitude ratios A_2/A_3 and A_1/A_3 from which the K_3/K_2 and K_3/K_1 ratios can be obtained and confirmed.

105

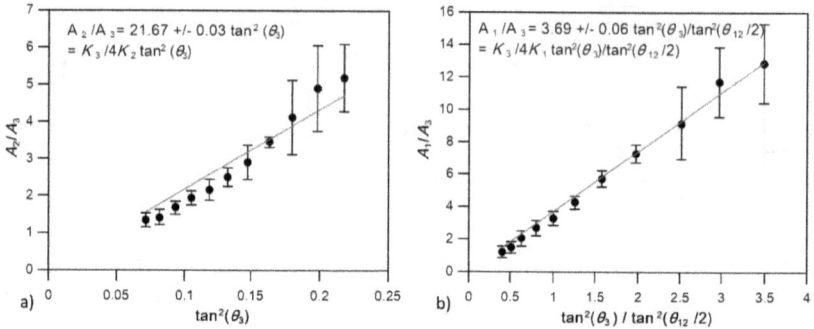

Figure 4.12: (a) Ratio of the twist over the bend amplitude A_2/A_3 graphed versus $\tan^2(\theta_3)$. From the fitting, the $K_3/4K_2$ ratio is estimated to be around 21.67 ± 0.03. (b) Ratio of the splay over the bend amplitude A_1/A_3 graphed versus $\tan^2(\theta_3)/\tan^2(\theta_{12}/2)$. From the fitting, the $K_3/4K_1$ ratio is estimated to be around 3.69 ± 0.06. θ_3 is the scattering angle in the bend geometry, whereas θ_{12} represents the scattering angle in the splay-twist geometry.

The three relative ratios of the elastic constants obtained from the fits in Figure 4.11d and Figure 4.12 are the following: $K_2 : K_1 : K_3 = 1 : 5.87 : 86.7$. These numbers confirm the absolute values for the elastic constants for the investigated LLC obtained by calibration with 5CB, which are listed in Table 4.2 and compared to the literature data for thermotropic 5CB, lyotropic polymeric and lyotropic chromonic LCs.[56,67,68,115,116,118–120] Note that when taking the refraction between the index matching fluid and the LLC into account, a more complex analysis involving the different refractive indices has to be performed, see appendix. This was done too, and more precise elastic constants were obtained, shown in Table 4.2 in the last line. Note that mostly the bend elastic constant is affected by taking that correction into account.

Table 4.2: Summary of the measured elastic constants of the micellar lyotropic nematic LC CDEAB/DOH/H$_2$O compared to what is known from literature for splay (K_1), twist (K_2) and bend (K_3) elastic constant values for thermotropic 5CB and lyotropic systems. If not stated otherwise, measurements were performed at room temperature.

Nematic LCs	K_1 /pN	K_2 /pN	K_3 /pN
Thermotropic 5CB[119]	6	3	8
Lyotropic micellar N$_D$ CsPFO/H$_2$O at 309 K[118]	≈ 2	-	≈ 8
Lyotropic polymeric N$_C$ PBG[116,120]	4.1±0.2	0.36±0.02	4.7±0.3
Lyotropic chromonic N$_C$ Disodium cromoglycate DSCG[68,115]	10±1	0.7±0.1	24±3
Lyotropic chromonic N$_C$ Sunset yellow[67,68]	4.3±0.4	0.7±0.07	6.1±0.6
Lyotropic micellar N$_D$ CEDAB/DOH/H$_2$O:	1.98±0.17	0.34±0.03	25.7±2.1
Refraction-corrected values:	1.79±0.07	0.33±0.01	20.8±1.0

The splay constant of the investigated lyotropic LC is in the same range as the splay constants of thermotropic nematic 5CB (6 pN), chromonic lyotropic nematic SSY (4.3 pN), polymeric lyotropic nematic PBG (4.1 pN) or lyotropic micellar nematic CsPFO/H$_2$O (≈ 2 pN).

More surprisingly, the twist elastic constant of the investigated micellar LLC is roughly one order of magnitude smaller than the splay constant (1.79 pN) and even roughly two orders of magnitude smaller than the bend constant (20.8 pN). As already mentioned before, the relatively high bend elastic constant of the N$_D$ phase of the CDEAB/DOH/H$_2$O system can be explained by the nearby more ordered lamellar phase, which only allows for splay deformation and therefore the bend elastic constant diverges in its vicinity, see phase diagram in Figure 4.8d. In chromonic LLCs the nearby more ordered phase is a columnar phase that excludes splay and twist deformations.

Therefore, the conclusion can be drawn, that the K_1/K_3 ratio of chromonic LLCs, having a columnar phase as the nearby more ordered phase, is higher compared to the K_1/K_3 ratio of the investigated micellar LLC which has a lamellar phase as the nearby more ordered phase.

Regardless of that, both systems adopt mirror symmetry broken configurations under capillary confinement, because the twist elastic constant is very small in both systems. Compared to the literature known K_2 values for lyotropic chromonic and polymeric LCs with $K_2 = 0.36 - 0.7$ pN, the K_2 of the N_D of the micellar LLC is at the lower end being in the range of K_2 of the N_C of the polymeric PBG LLC. Therefore, the light scattering results confirm that the occurrence of reflection symmetry broken configurations of the micellar LLC CDEAB/DOH/H$_2$O under capillary confinement can be explained by a small twist elastic constant – similar to the case of chromonics. The system tends to release energetically costly splay and bend deformations by twisting into helical structures. Additionally, the small twist elastic constant confirms the stability of the twist disclinations in the twisted polar configuration, as $K_2 < \frac{K_1+K_3}{2}$,[88,89] see Chapter 3.4.1.2. This phenomenon of a small twist elastic constant – and therefore leading to reflection symmetry broken configurations – does not seem to be restricted to the special case of chromonic LLCs; it seems to be a more general phenomenon in lyotropic LCs.

The physics behind why a small twist elastic constant seems to be a special feature of lyotropic LCs is still to be investigated. According to de Gennes[14], the elastic constants of LLCs should be in general smaller than the ones of thermotropic LCs, because the elastic constants decrease with increasing size of mesogenic building blocks. Thus, micellar, chromonic and polymeric nematics are expected to have smaller elastic constants than thermotropic nematics in which the building blocks are single organic molecules with anisotropic shape. In fact, no mirror symmetry broken configurations of achiral thermotropic LCs under capillary confinement have been reported in similar studies using, e.g., MBBA or 5CB. Anomalies in the elastic moduli originating from the large supramolecular building blocks of achiral LLCs might be a reason for the mirror symmetry breaking under achiral confinement conditions but why the twist elastic modulus is so much smaller than the other two moduli cannot be simply explained by that.

The main difference between thermotropic and lyotropic LCs is that the lyotropic mesogenic building blocks are surrounded by a solvent whereas thermotropic LCs by definition do not contain any solvent. One might speculate that the solvent can serve as a kind of lubricant.

In 2018, Hata et al. found that the twist elastic constant B_3 in a thermotropic smectic C phase can decrease monotonically up to an order of magnitude by swelling the LC with a solvent.[126] From this softening of twist elasticity, it was concluded that the solvent prevents direct collisions between LC molecules from neighboring layers and weakens the steric interaction which is thought to govern the interlayer orientation order across the layers. But nevertheless, the SmC phase is stable even in the range of a solvent volume fraction up to 2.5. By measuring also the rotational viscosity coefficient γ_1 independently, and calculating the dispersion relation B_3/γ_1, it was stated that the intercalated solvent layers in the swollen SmC phase provide lubrication with respect to γ_1 and as a result, the twist elastic constant B_3 of the swollen SmC phase becomes one order of magnitude smaller than in the original non-swollen SmC phase.[126] This reasoning can be applied to lyotropic LCs in which the solvent volume fraction is definitely higher than in the example of the swollen SmC phase. Furthermore, LLC systems are much more flexible and dynamic, since the micelle formation process is in constant equilibrium with surfactant molecules continuously exchanging between a micelle and the surrounding solvent and then incorporating into neighboring micelles and so on.

Zhou et al. discussed the smallness of K_2 also under the aspect of the flexibility of the aggregates for the case of the lyotropic chromonic N_C phase of DSCG.[68] They expanded the theoretical Onsager type modeling of rigid long rods by introducing flexibility as a finite persistence length λ_p. It was found, that the flexibility influences the bend elastic constant since bend deformation is no longer inhibited by the length of the aggregates as each aggregate can bend to follow the director pattern. Whereas the splay constant is not affected significantly by a λ_p because splay deformations under the condition of constant density limit the freedom of molecular ends which decreases the entropy, see Figure 6 in ref.[68]. Finally, the twist elastic constant in the model of flexible rods was implied to have a rather weak dependence on the flexibility and the volume fraction. Zhou et al.[68] argued that even though the aggregates point in different directions in a twisted structure, they can still diffuse along the local twist axis and encounter building blocks with different orientations, as thermal fluctuations can displace the aggregates and therefore cause an interaction between aggregates of different orientation. But whether this can cause the one order of magnitude decrease of K_2 in lyotropic nematics is still obscure.

4.3.4 Splay, twist and bend viscosities

Finally, the corresponding viscosities of the splay, twist and bend deformations η_{splay}, η_{twist} and η_{bend} are calculated, according to Equations (26), (27) and (36). In Figure 4.13 the inverses of the corresponding relaxation times $T_{eff,i}$ are graphed versus $q(\theta)^2$ for the splay and twist viscosity and versus $q(\theta)^2 \cos^2(\theta/2)$ for the bend viscosity. Note that the refraction-corrected θ and q values were used, see Equations (60 – 61) and (64 – 65) in the appendix.

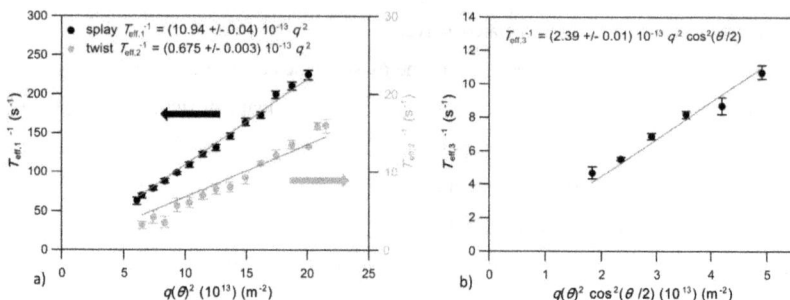

Figure 4.13: (a) Inverse relaxation times $T_{eff,1}$ and $T_{eff,2}$ of the splay (black dots) and twist (green dots) mode respectively plotted against $q(\theta)^2$. The slopes correspond to the ratios K_1/η_{splay} and K_2/η_{twist} respectively. (b) Inverse relaxation times $T_{eff,3}$ of the bend mode plotted versus $q(\theta)^2\cos^2(\theta/2)$. The slope $(2.39\pm0.01)\ 10^{-25}\ m^2s^{-1}$ corresponds to the ratio K_3/η_{bend}.

The fitting of the linear functions of Figure 4.13a gives a slope of $K_1/\eta_{splay} = (10.94\pm0.04)\ 10^{-25}\ m^2s^{-1}$ for the splay mode and a slope of $K_2/\eta_{twist} = (0.675\pm0.003)\ 10^{-25}\ m^2s^{-1}$ for the twist mode.

Inserting the splay and twist elastic constant $K_1 = 1.79$ pN and $K_2 = 0.33$ pN respectively gives a splay viscosity of $\eta_{splay} = (1.64\pm0.07)$ kgm^{-1}s^{-1} and a twist viscosity of $\eta_{twist} = (4.88\pm0.1)$ kgm^{-1}s^{-1} for the N$_D$ phase of the investigated LLC CDEAB/DOH/H$_2$O.

Fitting the linear function in Figure 4.13b gives a slope of $K_3/\eta_{bend} = (2.39\pm0.01)\ 10^{-25}\ m^2s^{-1}$ for the bend mode. Inserting the bend elastic constant of $K_3 = 20.8$ pN gives a bend viscosity of $\eta_{bend} = (87\pm4)$ kgm^{-1}s^{-1}.

Table 4.3 summarizes the measured splay, twist and bend viscosities of the micellar lyotropic CDEAB/DOH/H$_2$O N$_D$ phase in comparison to known viscosity values for the N$_C$ phases of thermotropic 5CB, a lyotropic polymeric LC and a lyotropic chromonic LC.

110

Table 4.3: Summary of the viscosities of the splay, twist and bend viscosities η_{splay}, η_{twist} and η_{bend} measured for the micellar lyotropic CDEAB/DOH/H$_2$O N$_D$ phase at room temperature compared to what is known from the literature for the thermotropic N$_C$ phase of 5CB and other lyotropic systems at room temperature.

Nematic LCs	η_{splay} /kgm^{-1}s^{-1}	η_{twist} /kgm^{-1}s^{-1}	η_{bend} /kgm^{-1}s^{-1}
Thermotropic N$_C$ 5CB[119]	0.078	0.08	0.028
Lyotropic polymeric N$_C$ PBG[116,120]	3.5	3.5	0.016
Lyotropic chromonic N$_C$ Disodium cromoglycate[68,115]	25	20	0.013
Lyotropic micellar N$_D$ CEDAB/DOH/H$_2$O	1.64±0.07	4.88±0.1	87±4

What attracts attention is that in all cases of a N$_C$ phase, η_{bend} is the smallest viscosity compared to the splay and twist viscosities, which is especially distinctive in the lyotropic LCs PBG and DSCG. This effect can be explained according to the Ericksen-Leslie model with the backflow mechanism.[127-130] The splay, twist and bend viscosities for nematics in terms of the Leslie coefficients and Miesowicz viscosities can be described as:[127-131]

$$\eta_{splay} = \gamma_1 - \frac{\alpha_3^2}{\eta_1} \tag{49}$$

$$\eta_{twist} = \gamma_1 = \alpha_3 - \alpha_2 \tag{50}$$

$$\eta_{bend} = \gamma_1 - \frac{\alpha_2^2}{\eta_2} \tag{51}$$

with the rotational viscosity coefficient γ_1, two of the six Leslie coefficients namely α_2 and α_3, the viscosities η_1 corresponding to **n** parallel to the flow direction, η_2 corresponding to **n** parallel to the gradient of flow and η_3 corresponding to **n** perpendicular to the flow direction as well as to the gradient of flow.[127-131] The rotational viscosity γ_1 is also often called twist viscosity; it generally determines the rate of relaxation of the director. The difference $\eta_1 - \eta_2$ is named the torsion coefficient and characterizes the contribution to the torque and reflects a coupling between the internal dissipation and the shear flow. The viscosity η_3 (not mentioned here) corresponds to the usual dynamic viscosity that arises in standard isotropic Newtonian fluids.[127–131] Figure **4.14**a-c illustrates the three Miesowicz viscosities for the case of calamitic shaped building blocks.

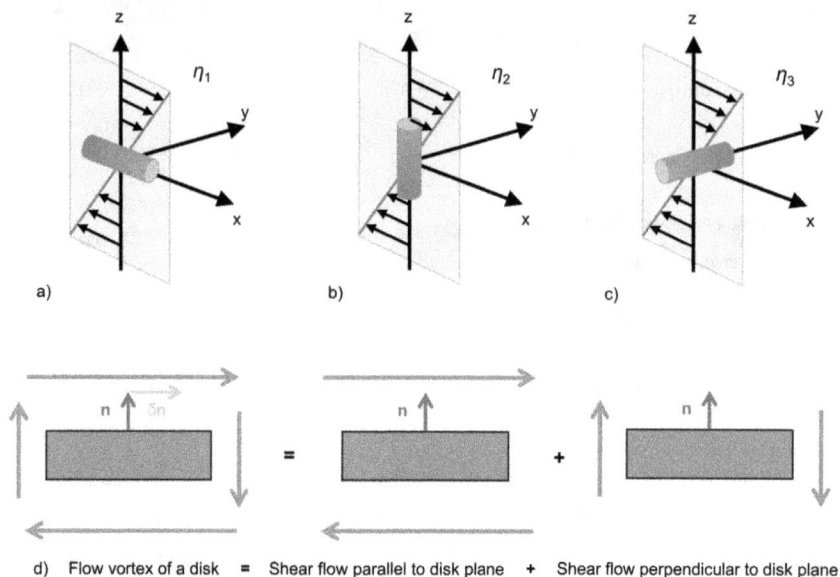

a)   b)   c)

d) Flow vortex of a disk = Shear flow parallel to disk plane + Shear flow perpendicular to disk plane

Figure 4.14: Schematic illustration of the three Miesowicz viscosities in the case of calamitic shaped building blocks. (a) η_1 corresponding to when **n** is parallel to the flow direction. (b) η_2 corresponding to **n** parallel to the gradient of flow and (c) η_3 corresponding to **n** perpendicular to the flow direction as well as to the gradient of flow. (d) Schematic illustration of how the flow vortex of a rotating disc can be decomposed into two pure shear flows, one parallel and one perpendicular to the disc plane. The director **n** is marked as a red arrow. A small deviation of the director orientation δ**n** (yellow arrow) perpendicular to **n** evokes a rotation of the disk. Sketch (a) - (c) is redrawn from ref. [130]. Sketch (d) is redrawn based on Figure 3 in ref. [129].

Typically, this is discussed for the case of calamitic mesogens, but because the investigated LLC consists of disc-shaped mesogens, the following discussion will transfer the backflow mechanism from calamitic to discotic building blocks, as described in ref. [127–129].

The flow vortex of a rotating disk can be decomposed into a shear flow which is parallel and one that is perpendicular to the disk plane, see Figure 4.14d. In the case of a splay deformation, the shear flows parallel to the disks cancel each other out, whereas the shear flows perpendicular to the disk plane (meaning parallel to the director **n**) give a constructive and therefore macroscopic backflow. The backflow tends to suppress relative motions of the micelles. The internal dissipation is decreased and only some residual dissipation from the local shear parallel to the disk plane can occur. Thus, a reduction of the orientational viscosity due to backflow is usually large for splay deformations in the case of an N_D phase.

In the case of bend fluctuations, the shear flows which are perpendicular to the disk plane cancel out whereas those parallel to it interfere constructively suppressing the friction from shear along that direction, but this gives no increase of internal dissipation. Hence, in the N_D phase, the coupling between the order and the flow is smaller for bend fluctuations and stronger for splay fluctuations which then give a stronger reduction of the splay viscosity, thus a smaller η_{splay} value.

For a N_C phase this behavior is opposite and it is the bend viscosity which gets strongly reduced by the backflow effect.[127–129] This can be explained by the fact that from a two-dimensional side view, a splay deformation of N_D looks like a bend deformation in a N_C phase – or in other words, a bend deformation of a N_D looks like a splay deformation in a N_C phase, see Figure 1.3.

In the N_C as well as in the N_D, twist fluctuations would give no backflow, as the interference is totally destructive and therefore $\eta_{twist} = \gamma_1$, see Equation (50).[127–129] For rod-shaped mesogens, see Figure 4.14a-c, the viscosity η_1 – corresponding to when **n** is parallel to the flow direction – would be smaller compared to the viscosity η_2 (where the director **n** is perpendicular to the flow) because if **n** is parallel to the flow, the rods can slide along each other. This is opposite for disc-shaped mesogens; here the viscosity η_2 is smaller than η_1, compare Equations (49) and (51). So, the normal case of a N_C phase as in thermotropic 5CB, see Table 3.3, would be that η_{twist} is of the highest, closely followed by η_{splay} and η_{bend} is of the smallest.[127–129]

Table 4.3 shows that in all investigated N_C LLCs the splay and twist viscosities are more or less in the same range or even equal to each other within the accuracy of the experiment, whereas the bend viscosity is of two orders of magnitude smaller compared to η_{splay} and η_{twist}. The splay and twist viscosities of the CDEAB/DOH/H_2O system are in a similar range compared to each other and compared to η_{splay} and η_{twist} of the lyotropic polymeric N_C PBG. The fact that the twist viscosity is higher than the splay viscosity is in agreement with the theory of the Leslie coefficients and Miesowicz viscosities.

But – whereas η_{bend} of the lyotropic polymeric N_C PBG and the lyotropic chromonic N_C Disodium cromoglycate are two to three orders of magnitude smaller than the corresponding η_{splay} and η_{twist} – the bend viscosity of the investigated micellar N_D LLC is two orders of magnitude higher than the respective η_{splay} and η_{twist}, which is in contradiction to Equations (49 – 51) where the twist viscosity is set to be highest in the Ericksen-Leslie model. In the case of DSCG, similar anomalous behavior can be observed since $\eta_{\text{splay}}/\eta_{\text{twist}} \approx 1$ only holds in the vicinity of the transition to the isotropic phase. Being deeper in the nematic phase, the ratio can be as large as 2, meaning that – like perhaps in the CDEAB/DOH/H_2O system – η_{twist} can decrease significantly due to an anomalously small K_2.[68]

Furthermore, the difference in the bend and twist elastic constant of the micellar LLC is much greater than the difference in K_1 and K_2 of DSCG, meaning that this effect could be much more pronounced in the investigated micellar LLC. As the chromonic DSCG system consists of rod-shaped mesogens, and the micellar CDEAB/DOH/H_2O consists of disc-shaped micelles, a reversed behavior with respect to the backflow mechanism for η_{splay} and η_{bend} is to be expected – which seems to be the case. Additionally, the high η_{bend} of the micellar LLC could be explained by the proximity to nearby more ordered lamellar phase. The disk-shaped micelles can undergo strong lamellar fluctuations, which inhibit bend deformations and could cause an increase in the bend elastic constant, as well as in the bend viscosity.

The anomalous behavior of the η_{twist} is still to be understood. Zhou et al.[68] suggest that an explanation for this is rooted in the flexibility of the mesogens and the strong temperature and concentration dependencies of the mean aggregate length (or in our case width). This is similar to the anomalous peculiarity of the twist elastic constant, in which an overall much more dynamic system and a solvent serving as lubrication could generate a decreased twist elastic constant, compared to the splay and bend constants.

4.4 Chapter Conclusion

In conclusion, this chapter shows the results of a light scattering study which gives experimental evidence that the twist elastic constant of the investigated nematic phase of the micellar LLC CDEAB/DOH/H_2O is one or even two orders of magnitude smaller than the splay and bend elastic constants. This finding explains – as in the case of chromonic LLCs[61,68] – the occurrence of reflection symmetry broken configurations under capillary confinement. Therefore, a small twist modulus is not just a special feature of chromonic LLCs; it seems to be a much more general phenomenon in lyotropic liquid crystals, including all kinds of lyotropic systems ranging from standard micellar to chromonic and to polymeric LLCs. A theoretical explanation of that phenomenon is still to be found, but one might speculate – comparing lyotropic LCs to thermotropic LCs – that the solvent in LLCs could serve as lubrication. Furthermore, the flexibility of the mesogens and the much more dynamic behavior of lyotropic systems have to be considered.

4.5 Appendix

The formalism of the light scattering analysis on nematic LCs in Chapter 4.2.3.2 was elaborated in detail by Peter Collings. This appendix will focus on the origin of the two scattering geometries – the splay-twist and the twist-bend – and the corresponding analysis. Motions of LC molecules or building blocks range over a wide frequency range. For example, the rotational and transitional motion of a single molecule is very fast (10^{-9} to 10^{-11}s), or density fluctuations due to molecular translational mobility, which are also present in the isotropic phase. But those scatter light very weakly, whereas in LCs spatially correlated collective orientational fluctuations, which appear at lower frequencies, are distinctive for e.g. a liquid crystalline nematic phase. They include the director fluctuations which describe local fluctuations in the orientation of the director, and the fluctuations of the order parameter S. In nematic LCs, the director fluctuations are more important than the order parameter fluctuations because they are easier to excite. The order parameter fluctuations only come into play near the nematic to isotropic phase transition.[14,121] In the following, the thermal fluctuations of the director in the bulk nematic LC are discussed.

Considering the fluctuational part in the Frank free energy and analyzing the fluctuation quantities in terms of planar waves, it is reasonable to introduce a new coordinate system with the average director orientation in the nematic sample along the z-axis. The two unit vectors e_1 and e_2 are chosen so that they are perpendicular to the z-axis and to each other. The e_2 unit vector is set perpendicular to the scattering vector q. Furthermore, the fluctuation eigenmodes $n_\alpha(q)$ are along e_α ($\alpha = 1, 2$). The fluctuation eigenmode $n_1(q)$ describes a combination of splay and bend fluctuations, whereas the eigenmode $n_2(q)$ is a combination of twist and bend fluctuations. Expressing the scattering intensity in terms of the two eigenmodes, the following equation is the basis for the differential cross-section per unit solid angle of the outgoing beam:[14,113,121]

$$I = X' \left(\frac{\Delta\varepsilon\,\pi}{\lambda_0^2}\right)^2 d\,P \left[\frac{(i_1 f_z + i_z f_1)^2}{K_1 q_\perp^2 + K_3 q_\parallel^2} + \frac{(i_2 f_z + i_z f_2)^2}{K_2 q_\perp^2 + K_3 q_\parallel^2}\right], \tag{52}$$

with λ_0 as the vacuum wavelength of light, $\Delta\varepsilon$ as dielectric anisotropy of the sample, d as the sample thickness, P as the laser power, i and f as incident and scattered polarizations, q as the scattering wave vector and $q_\parallel$ and $q_\perp$ being the components of q parallel and perpendicular to the director **n** respectively, and a calibration factor X' including some constant factors like the thermal energy $k_B T$ and optical parameters.

Applying this formula to the geometrical situations of the splay-twist and the twist-bend geometry, the scheme that is shown in Figure 4.15 illustrates the two cases.

Splay - Twist Geometry

a)

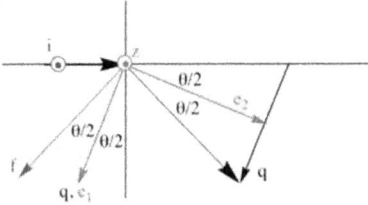

$i_1 = 0$
$i_2 = 0$
$i_z = 1$
$f_1 = \cos(\Theta/2)$
$f_2 = \sin(\Theta/2)$
$f_z = 0$
$q_{||} = 0$
$q_\perp = 1$

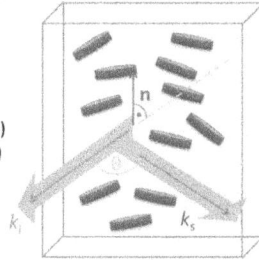

Twist - Bend Geometry

b)

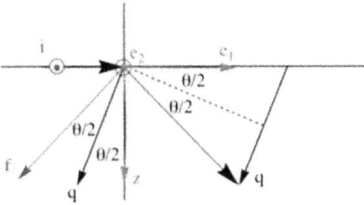

$i_1 = 0$
$i_2 = 1$
$i_z = 0$
$f_1 = \sin(\Theta)$
$f_2 = 0$
$f_z = \cos(\Theta)$
$q_{||} = \cos(\Theta/2)$
$q_\perp = \sin(\Theta/2)$

k_i = wave vector incident light
k_s = wave vector scattered light

Figure 4.15: (a) Geometrical situation of the splay-twist geometry. On the left side the scenario in the coordinate system which has the average director orientation in the nematic sample along z, and the two unit vectors e_1 and e_2, showing the incoming polarizations i and the outgoing scattered polarizations f and the components of q parallel and perpendicular to the director **n** ($q_{||}$ and $q_\perp$) respectively is demonstrated. On the right side, the corresponding situation in a sample cell with the wave vector k_i of the incident light and the wave vector k_s of the scattered light is illustrated. (b) shows the equivalent situations for the twist-bend geometry.

Applying the geometrical conditions of the splay-twist geometry, see Figure 4.15a, simplifies Equation (52) for the splay-twist scattering intensity I_{12} according to:

$$I_{12} = X' \left(\frac{\Delta \varepsilon \, \pi}{\lambda_0^2}\right)^2 d \, P \left[\frac{\cos^2\left(\frac{\theta}{2}\right)}{K_1 q^2} + \frac{\sin^2\left(\frac{\theta}{2}\right)}{K_2 q^2}\right]. \qquad (53)$$

Applying the geometrical conditions of the twist-bend geometry, see Figure 4.15b, simplifies Equation (52) for the twist-bend scattering intensity I_{23} according to:

$$I_{23} = X' \left(\frac{\Delta \varepsilon \, \pi}{\lambda_0^2}\right)^2 d \, P \left[\frac{\cos^2(\theta)}{K_2 q^2 \sin^2\left(\frac{\theta}{2}\right) + K_3 q^2 \cos^2\left(\frac{\theta}{2}\right)}\right]. \qquad (54)$$

Applying the q angular dependence of the intensity of the twist mode $I_2 \sim \sin^2(\theta/2)$ gives the following equations:

$$I_{12} = X' \left(\frac{\Delta \varepsilon^2}{16 \lambda_0^2 n^2}\right) d \, P \left[\frac{1}{K_1} \cotan^2\left(\frac{\theta}{2}\right) + \frac{1}{K_2}\right] \qquad (55)$$

$$I_{23} = X' \left(\frac{\Delta \varepsilon^2}{16 \lambda_0^2 n^2}\right) d \, P \left[\frac{\cos^2(\theta)}{K_2 \sin^4\left(\frac{\theta}{2}\right) + K_3 \sin^2\left(\frac{\theta}{2}\right) \cos^2\left(\frac{\theta}{2}\right)}\right]. \qquad (56)$$

Note, that X' in Equations (52 – 56) collects some more constant factors and becoming X in Equations (30), (37) and (44).

If the twist elastic constant is assumed to be small compared to the bend elastic constant:

$$K_2 \sin^2\left(\frac{\theta}{2}\right) \ll K_3 \cos^2\left(\frac{\theta}{2}\right) \qquad (57)$$

and the (twist-)bend scattering intensity I_3 can be approximated to:

$$I_3 \approx X' \left(\frac{\Delta \varepsilon^2}{16 \lambda_0^2 n^2}\right) d \, P \left[\frac{4 \cotan^2(\theta)}{K_3}\right]. \qquad (58)$$

This approximation might work for lyotropic LCs, but not for thermotropic LCs, like 5CB.

Another correction, which has to be taken into account, is due to refraction between the index matching fluid and the sample, involving the different refractive indices n_L, $n_{\parallel}$ and $n_{\perp}$ which correspond to the refractive index of the index matching fluid, and the two refractive indices of the LC sample polarized along and perpendicular to the director **n**, respectively. First, there is no refraction for the incident beam because the angle to the normal of the interface is zero.

Second, the glass around the sample does not affect the angle since the beam goes in and out of the glass. Third, the angle in the index matching fluid is the angle measured in the lab since the matching fluid container is cylindrical (no refraction at the wall).

There are two possibilities for how to implement the birefringence correction – either change the theory so it is in terms of the lab angle θ_L, or change the experimental results so they are in terms of the sample angle θ, which is done in the following.

The conversion for θ_L to θ is different for the two geometries, see Figure 4.16. The refractive index for decalin[3] (Decahydronaphthalene) which was used as index matching fluid is $n_L =$ 1.481.[132]

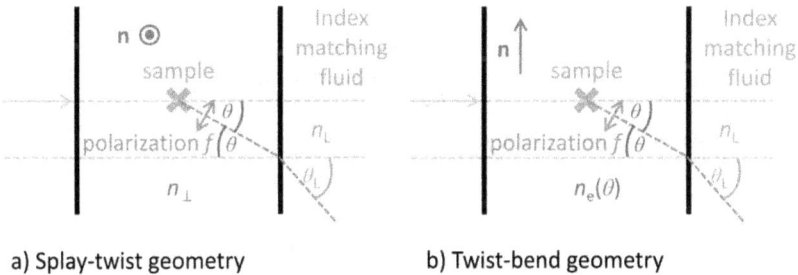

a) Splay-twist geometry   b) Twist-bend geometry

Figure 4.16: (a) In the splay-twist geometry, the output polarization f is perpendicular to the director **n**. The relevant refractive index of the sample is $n_\perp$ and the refractive index of the index matching fluid is n_L. θ_L is the lab angle, whereas θ is the sample angle. (b) In the twist-bend geometry the output polarization f makes an angle θ with the director **n**. The relevant refractive index of the sample is the angular dependent effective refractive index $n_e(\theta)$.

Let us discuss the splay-twist geometry first. According to Snell´s law, it is:

$$n_\perp \sin(\theta) = n_L \sin(\theta_L) \ . \tag{59}$$

For the sample angle θ follows:

$$\theta = \sin^{-1}\left(\frac{n_L \sin\theta_L}{n_\perp}\right) \tag{60}$$

---

[3] Mixture of cis- and trans-decalin

and the corrected q is:

$$q = \frac{2\pi}{\lambda_0} \sqrt{n_{\parallel}^2 + n_{\perp}^2 - 2n_{\parallel} n_{\perp} \cos(\theta)} \ . \tag{61}$$

The splay-twist intensity I_{12} divided by the laser power P under the assumption that the cell thickness d is constant can then be written as:

$$\frac{I_{12}}{P} = X \left(\frac{n_{\parallel}^2 - n_{\perp}^2}{q}\right)^2 \left[\frac{\cos^2\left(\frac{\theta}{2}\right)}{K_1} + \frac{\sin^2\left(\frac{\theta}{2}\right)}{K_2}\right] \ . \tag{62}$$

In the twist-bend geometry, the extraordinary refractive index $n_e(\theta)$ which is angular dependent has to be introduced:

$$n_e(\theta) = \frac{n_{\parallel} n_{\perp}}{\sqrt{n_{\parallel}^2 \sin^2(\theta) + n_{\perp}^2 \cos^2(\theta)}} \ . \tag{63}$$

Applying again Snell´s law, for the sample angle θ results in:

$$\theta = \tan^{-1} \sqrt{\frac{n_{\perp}^2 n_{\parallel}^2 \sin^2(\theta_L)}{n_{\parallel}^2 (n_{\perp}^2 - n_{\perp}^2 \sin^2(\theta_L))}} \ , \tag{64}$$

and for the q vector follows:

$$q = \frac{2\pi}{\lambda_0} \sqrt{n_e^2 + n_{\perp}^2 - 2n_e n_{\perp} \cos(\theta)} \ . \tag{65}$$

The twist-bend intensity I_{23} divided by the laser power P under the assumption that the cell thickness d is constant, can be then written as:

$$\frac{I_{23}}{P} = X \left(\frac{n_{\parallel}^2 - n_{\perp}^2}{q}\right)^2 \left[\frac{\cos^2(\theta)}{K_2 \sin^2\left(\frac{\theta}{2}\right) + K_3 \cos^2\left(\frac{\theta}{2}\right)}\right] \ . \tag{66}$$

For the 5CB calibration, for each measurement angle θ – in the splay-twist as well as in the twist-bend geometry – the factor X_{5CB} is calculated. These X_{5CB} are averaged to get a mean calibration factor $\overline{X}_{5CB}$.

The lyotropic elastic constants can then be obtained from the following equations:

$$K_1^{\text{lyo}} = \overline{X}_{5\text{CB}} \left(\frac{n_\parallel^2 - n_\perp^2}{q}\right)^2 \left(\frac{\cos^2\left(\frac{\theta}{2}\right)}{\left(\frac{I_1^{\text{lyo}}}{P}\right)}\right) , \qquad (67)$$

$$K_2^{\text{lyo}} = \overline{X}_{5\text{CB}} \left(\frac{n_\parallel^2 - n_\perp^2}{q}\right)^2 \left(\frac{\sin^2\left(\frac{\theta}{2}\right)}{\left(\frac{I_2^{\text{lyo}}}{P}\right)}\right) , \qquad (68)$$

$$K_3^{\text{lyo}} = \overline{X}_{5\text{CB}} \left(\frac{n_\parallel^2 - n_\perp^2}{q}\right)^2 \left(\frac{\cos^2(\theta)}{\left(\frac{I_3^{\text{lyo}}}{P}\right) \cos^2\left(\frac{\theta}{2}\right)}\right) . \qquad (69)$$

In the splay-twist geometry a splay and twist elastic constant K_1^{lyo} and K_2^{lyo} can be calculated for each measurement angle θ. Those values are being averaged respectively. In the bend geometry the bend elastic constant K_3^{lyo} can be calculated for each measurement angle θ and the obtained values are averaged. The results are listed in Table 4.2.

5 Chirality detection using thermotropic nematic liquid crystal droplets on anisotropic surfaces

Chapter Overview

The previous chapters studied a lyotropic system with a peculiar anomalous small twist elastic modulus exhibiting mirror symmetry broken configurations under capillary confinement which can be exploited to measure qualitatively and quantitatively tiny amounts of chiral additives. Unfortunately, those chiral configurations cannot occur in thermotropic LCs because their elastic constants are typically within the same order of magnitude.[119] But, as an advantage compared to lyotropic LCs, thermotropic LCs can be exposed to air without any concerns. This opens new possibilities concerning the use of other confined geometries involving the air-LC interface as additional boundary condition. Like, e.g., the confinement of a sessile LC droplet on an anisotropic surface is used in the studies of the following chapter. By means of the two given boundary conditions at the air-LC and the LC-glass interfaces, a locally twisted hybrid director structure with a disclination line across the droplet is obtained. The shape of the disclination line (S- or 2-like) directly indicates the handedness of an induced chiral nematic phase and large pitch values up to 10 – 20 mm can be easily measured. This part of my thesis is essentially based on my publication with Per Rudquist, Andrew Mark and Frank Gießelmann on "Chirality Detection Using Nematic Liquid Crystal Droplets on Anisotropic Surfaces" which appeared 2016 in Langmuir.[133]

5.1 Motivation

In the first chapter the chiral nematic, also called cholesteric phase (N*), with its distinctive physical properties was already introduced, see page 15. In a chiral liquid crystalline phase, either the mesogens are chiral themselves or a chiral dopant is added to an achiral liquid crystalline host phase.[43,57,134-138] The helical superstructures in the N* or SmC* phase, induced by the chirality of added organic molecules are examples of transfer and amplification of chirality from a microscopic to a macroscopic scale. This transfer process is called chiral induction and can be described as an example of "amplification of information".[43,57,134-138] Liquid crystals can be regarded as a very sensitive analytical tool to study chirality and chirality effects. Although there have been empiric rules proposed,[135,138-146] the chirality transfer between the molecular structure of a chiral mesogen or an enantiomer acting as a chiral dopant and the respective macroscopic structure or response in the liquid crystalline phase is rather complicated and still not understood since the occurrence of chirality of a phase depends on many mechanisms with different weighting factors.[57,58,147,148] Chirality changes the physical properties of a liquid crystal and ferroelectric as well as chirally doped smectic A materials have been used for determination of enantiomeric excess in compounds that could be doped into such system.[149,150]

Chirality is a quality, not a quantity, and it is generally accepted that no universal measure of chirality exists. Therefore, any quantitative definition is related to a certain system and "chirality measurements" are based on e.g. circular dichroism, optical rotation, and *Helical Twisting Power HTP*.[43,57,134-138] The latter is only defined in liquid crystalline phases.[151] While the two enantiomers of a particular dopant, e.g. (*R*)- and (*S*)- mandelic acid, induce always the opposite sense of the helical superstructure in an, e.g., nematic phase,[152] one enantiomer can induce a right- or left-handed helical superstructure in different nematic hosts. The sign and magnitude of the induced helical pitch of liquid crystalline systems doped with type I dopants do only depend on the concentration of the dopant. Whereas, using type II dopants – which most dopants are – the sign and the magnitude of the pitch are depending on the concentration as well as on the used host material.[153,154] Therefore, having different host materials doped with the same enantiomer of a type II dopant lead to chiral phases of opposite handedness and different *Helical Twisting Power HTP*.

The method described in this chapter constitutes an easy and very sensitive way to test and analyze chirally doped nematic liquid crystals in terms of the induced handedness and the *HTP* of the dopant. Because the method is so sensitive, the focus lies on the detection of chiral induction from dopants at very small concentrations, e.g., down to 1 chiral dopant molecule per 1000 achiral host molecules (mole fraction of chiral dopant $x_d \approx 0.001$), and/or from dopants with extremely small *Helical Twisting Powers*. In these cases, the induced pitches P can be expected to be in the range of several tens of millimeters. The detection and determination of such weak chiral induction are related to the fundamental question whether there is a minimum concentration of chiral dopant for which chiral transfer can occur. First, typical methods known from literature to measure pitches in chiral nematic phases are summarized in the following.

The natural texture of an N* phase between two untreated glass substrates (with a cell thickness d) is the so-called oily streaks texture. The director **n** orients planar at the glass substrates which means that the helix axis of N* is going perpendicular to the glass plates. The oily streaks can be seen as a network of defect lines dispersed in uniformly helical regions. However, the structure of an oily streak is very complicated and depends mostly on the delicate interplay between elasticity of the LC and the surface anchoring. The oily streaks texture can develop with time to a uniformly oriented sample with the helix axis perpendicular to the glass substrates. This can be analyzed in transmittance polarizing microscopy by rotation of the sample between crossed polarizers. The pitch P of the N* can be roughly estimated if e.g. $d = N2\pi P$ meaning that several 360° turns of P fit into the given cell thickness d. This is the case if neither the observed color nor the transmitted intensity will change by rotation of the sample. When $d \approx P$ meaning that the pitch is in the range of the cell gap, elastic interactions deform the helix and the observed intensity can vary through rotation. If $2d < P$, meaning that twice the cell gap is not enough to fit in a single 360° turn of the helix, the extreme case of a uniform planar nematic texture can be observed.[21,147,155]

Cholesteric phases with a pitch smaller than 1 µm are referred to as short pitch materials. Hence, the phenomenon of selective light reflection can be used to determine the pitch. In this method, the De Vries relation between the wavelength of reflexion λ_0 and the pitch P with the mean birefringence $n = (n_\parallel + n_\perp)/2$ is applied:[21,147,155]

$$\lambda_0 = n P .$$

(70)

The wavelength of reflexion is the wavelength for which the incoming light is partially reflected by the helical structure. In the case of short pitch materials, the wavelength of selective reflexion lies in the visible spectrum of light, whereas in case of long pitch materials the wavelength is in the far-infrared. To observe a narrow reflection band, the material is filled into a cell with planar boundary conditions such that the helix of N* is oriented parallel to the direction of light which propagates perpendicular to the glass plates. The used cell thickness d should be adjusted such that at least 10 full 360° turns of the director fit into the gap. The linearly polarized incoming light beam consists of a left-handed circular polarized component and a right-handed component. In case of passing through a right-handed helix, the right-handed component of the incident light beam is being reflected whereas the left-handed component is transmitted. In the case of a left-handed helix, the behavior is vice versa. The wavelength can be measured by optical rotation dispersion, UV VIS spectroscopy or by measuring circular dichroism.[21,147,155]

In the case of homeotropic boundary conditions, the helix is oriented parallel to the glass plates, as it is shown in Figure 1.7 in the introductory chapter. Within the plane of the substrate, no preferred direction is given, so that the director of the twist helix is allowed to vary smoothly over macroscopic distances which give a bright-dark pattern. The equidistant dark stripes are due to when the local director is oriented along the direction of light propagation so at vanishing birefringence. The bright stripes appear when the local director is perpendicular to the direction of light propagation, so at maximum birefringence. The periodicity of the stripe pattern is given by $P/2$. If the cell thickness is becoming too small, approaching the range of P, the method isn´t accurate anymore because of deformations of the helix by elastic interactions with the confining glass substrates. In the case of even higher P, the texture is getting more complex due to more and more distorted director configurations like e.g. so-called cholesteric fingers or even pseudo-isotropic areas.[21,147,155]

This leads to the following methods for very long pitch materials, starting with the Grandjean-Cano geometries.[156,157] The Grandjean-Cano method is based on the observation of disclination lines which are due to a continuous change of the cell thickness to which the cholesteric material is exhibited. This continuous change in sample thickness can be realized by either using wedge geometry or by putting a lens on a glass substrate. A schematic example of the lens setup is shown in Figure 5.1. The basic idea is that a cholesteric LC under planar anchoring conditions shows disclination lines (Grandjean steps) whenever an integer multiple of the half-helix $P/2$ does not fit to the imposed cell gap.

Only helical structures with integer multiples N of the half-pitch $P/2$ are allowed and the regions with different values of N are separated by disclination lines.[57]

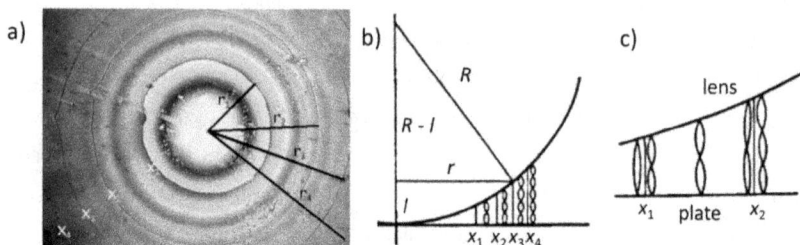

Figure 5.1: (a) Typical Grangjean texture of a cholesteric LC placed between a glass plate and a lens. (b) and (c) Cross-section of the setup between glass substrate and lens. The positions indicated as x_n mark the regions of an erratic change of the number of helical turns. In these regions the helical structure has to be compressed or stretched and disclination lines are formed. The radii of the Grandjean steps r are related to the lens radius R and the pitch P of the cholesteric phase. The photo and the scetch are reprinted with permission from ref. [158].

The Grandjean-Cano geometries are convenient for measuring pitch values of up to tens of micrometers but for pitches in the millimeter range, meaning a thousand times weaker twist, the dimensions of the wedge of lens become unpractical and other methods must be applied.[57]

Only very few methods for the determination of chiral nematic pitch values in the millimeter range have been reported so far in literature. Suh et al.[159] developed a method using circularly rubbed (CR) cells in which one of the alignment surfaces is uniformly rubbed along the 12 – 6 diameter (clock face notation), and the other inner glass surface is circularly rubbed, see Figure 5.2. Filling a CR cell with an achiral nematic LC, a planar twisted structure is adapted by the nematic director field. The twist varies continuously from 0° along the 9 – 3 diameter because here the upper and lower rubbing directions of the opposing inner glass surfaces are parallel to each other, to 90° along the 12 – 6 diameter where the rubbing directions are perpendicular to each other. Additionally, along the 12 – 6 diameter, the twist jumps from left-handed to right-handed in order to minimize the elastic energy. A disclination line separating the two oppositely twisted domains is formed.[159]

Filling a chiral LC into a CR cell, the degeneracy of left- and right-handed twist sense is lifted and the minimum elastic energy corresponds to a finite twist. This leads to an azimuthal rotation of the disclination line by an angle β away from the 12 – 6 diameter. The axis of lowest elastic energy density is always perpendicular to the disclination line. The pitch P is then given by the angle β and the cell thickness d:[159]

$$P = \frac{360° \, d}{\beta} \, . \tag{71}$$

To determine the possible limit of the CR method, let us say angles of down to 1° can be measured. For $\beta = 1°$ and $d = 20$ µm, a pitch value of 7.2 mm is obtained. So, pitch values of several millimeters can readily be measured. Analogously, the circular rubbed alignment surface can be exchanged for a radially "rubbed" surface, which can be achieved by means of photo buffing.[160] The methods based on the circularly as well as on the radially rubbed cell usage are straightforward and elegant. However, the cell preparation is rather complicated.

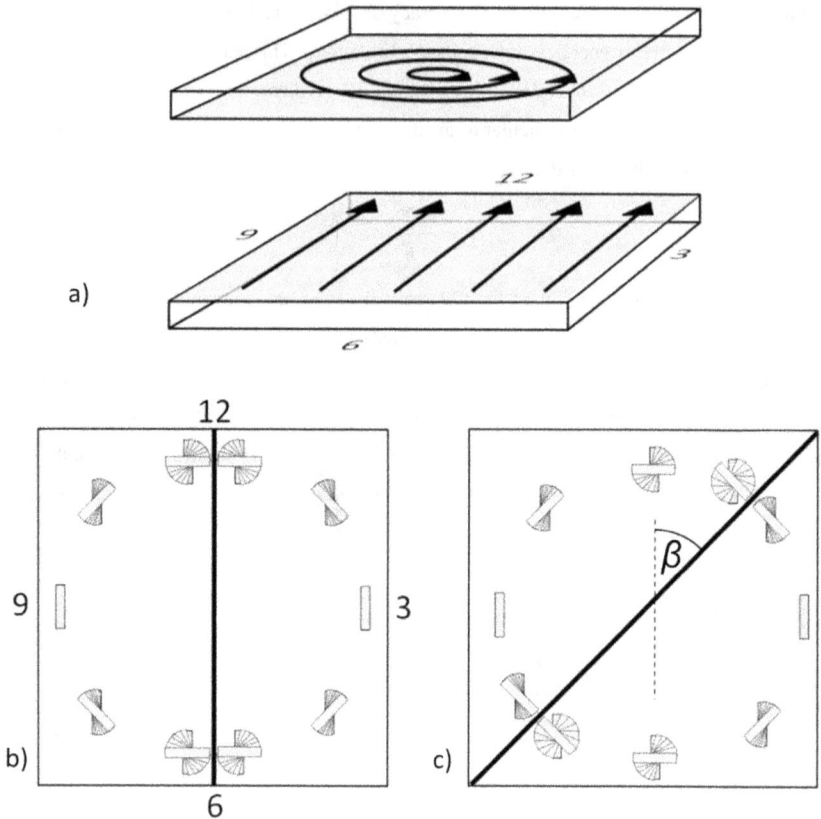

Figure 5.2: In the method proposed by Suh et al.[159], circularly rubbed (CR) cells are used. (a) One inner glass surface is unidirectionally rubbed, while the opposite surface is circularly rubbed. (b) Filling an achiral nematic LC into a CR cell, a disclination line is formed along the unidirectional rubbing direction. (c) Filling a chiral nematic LC into a CR cell, the helix sense and the pitch value are related to the rotation direction and the rotation angle β of the disclination line respectively. The axis of lowest elastic energy density is always orthogonal to the disclination line. Figure from ref. [133], reprinted with permission from Langmuir.

Even longer pitch values can be measured by the twisted nematic (TN) cell method developed by Raynes.[161,162] Here, one of the glass substrates is rubbed orthogonal to the rubbing direction of the other one, imposing a 90° twist on the nematic director field. In case of filling an achiral nematic LC into a TN cell, domains of left- and right-handed 90° twisted structures are formed and separated by disclination lines. These disclination lines can be pinned by particles, like the spacers used for securing the cell thickness.[162] In the achiral case, the disclination lines are perfectly straight. Whereas, when filling a chiral nematic LC into a TN cell the disclination lines get curved because the twist sense is biased by the addition of chirality and the domains of energetically favored twist sense grow at the expense of those with the opposite twist sense. The curvature radius R of the curved disclination lines is a measure of the pitch according to:[162]

$$P = 2R \ . \tag{72}$$

Pitch values up to 50 mm can be measured.[162] In the subsequent Chapter 5.3.2 a micrograph of a TN cell filled with achiral 5CB and 5CB doped with (S)-mandelic acid is shown as an example in Figure 5.9.

However, the limitations of this method are the following. As also discussed by Raynes,[162] the perfect parallel alignment of the nematic director at the rubbed glass surfaces is crucial. If the director makes a finite angle with the surface (surface pretilt), the degeneracy of left- and right-handed twist domains is lifted and therefore a finite pretilt would favor one twist handedness over the other when filling a nematic LC into the TN cell. This would result in curved disclination lines even if the LC material is achiral. Additionally, the angle between the rubbing directions when assembling the two glass plates of the TN cell can never be made exactly 90° for practical reasons. Any deviation from 90° gives a pretwist and lifts the degeneracy of left- and right-handed twist domains as well. Both – pretilt and pretwist – can compromise the measurements of very long pitch values by making the empty TN cells already inevitably chiral. A nonzero pretilt would also affect the results obtained by CR cells due to the same reasons. Conclusively, it is generally necessary to combine several different methods to rule out the effects of any occasional chiral disturbances that might falsify the results.

In this chapter, a surprisingly simple and still sensitive method for the determination of the sign and the magnitude of long pitch values is presented. The experimental equipment is just a flat glass substrate coated with a conventional planar aligning polymer layer on which a tiny amount – typically less than 0.01 μL – of the chiral nematic material is put on as a sessile droplet, and an optical microscope for analysis. The basic features of this new method are described and demonstrated by using 5CB as thermotropic nematic host material which is weakly doped with (R)- or (S)-mandelic acid. The helix sign of the chirally doped nematic is directly revealed by the shape of the appearing disclination line in the droplet.

The magnitude of the induced pitch can easily be measured as well. The underlying concept partly relates to studies of Yamaguchi and Satu[163] who investigated the features of achiral nematic LC droplets on aligning surfaces.

5.2 Materials and experimental techniques

The thermotropic "room-temperature nematic" host material 4-cyano-4´-pentylbiphenyl, known as 5CB, with clearing point at 35 °C was purchased from Sigma-Aldrich and Synthon and used without further purification. The chiral dopants (*R*)- and (*S*)-mandelic acid were purchased from Alfa Aesar and Sigma-Aldrich respectively. Mixtures of the nematic 5CB and the chiral dopant in the concentration range from 90 mmol% down to 2.7 mmol% were prepared using a stock solution and stirred in the isotropic phase. Each mixture was stirred again in a Thermoshaker at 40 °C before use in order to ensure the best possible homogenization. As substrate for the sessile droplets, commercially available LC cells (AWAT Spotka and EHC) coated with polyimide (NISSAN SE-130 and double side rubbed) were cracked open.

Very small sessile droplets of LC with volumes of typically less than 0.01 µL (footprint diameter $\approx$ 500 µm and droplet height of $\approx$ 100 µm) were deposited very carefully by means of an Eppendorf pipet tip on the polyimide coated glass substrate. Approximately 10 – 15 droplets with the same dopant concentration were assembled and analyzed with a polarizing optical microscope (Leica DM-LP), in combination with an INSTEC TS62 hot stage and a Nikon D40 camera.

The sessile droplets were heated into the isotropic phase to eliminate possible alignment effects from the deposition and subsequently cooled down into the nematic phase till room temperature. The forming disclination line needs some time to relax into its equilibrium configuration. Slowly cooling down with 0.5 K/min to room temperature turned out to be the best way to achieve reproducible and minor-defective results compared to processes using quenching and cooling down to just below the clearing point. After the sample reached room temperature, an additional half an hour waiting is enough to be sure that the equilibrium configuration of the disclination line is achieved.

The long relaxation times can be understood from the very small differences in elastic energy involved between domains of opposite twist sense on both sides of the disclination line, see Chapter 5.3. Similar long relaxation times of the disclination lines are required as well for Grandjean-Cano experiments and the CR cell method from Suh et al..[156,157,159]

For comparison, pitch values of the investigated chiral nematic mixtures were measured with the Raynes method in TN cells as well in order to have a reference. The TN cells were fabricated in the MC2 Cleanroom facility of the Chalmers University of Technology in Gothenburg. Clean 75 mm x 75 mm glass plates were spin-coated with the polyimide PI2610 (DuPont), cured at 300 °C for 3 h, and subsequently rubbed with a velvet cloth in a commercial rubbing machine. The glass substrates were glued together by UV-curing glue with dispersed silica spacers to define the cell thickness. Each assembly gives 25 identical TN cells by scribing and breaking. The cells were filled with the LC by capillary forces, heated into the isotropic phase and cooled down to the nematic phase at room temperature. The director field equilibrated within the same relaxation time as for the droplets.

5.3 Results and Discussion

5.3.1 Achiral and chiral nematic LC sessile droplets

Figure 5.3a shows a polarized optical micrograph of an achiral sessile droplet of pure 5CB put on an anisotropic surface with homogenous planar anchoring conditions. The LC-air interface has an approximately spherical shape and a single straight disclination line is running perpendicular to the rubbing direction. Figure 5.3b,c show weakly doped chiral nematic 5CB droplets with 0.018 mol% (*R*)-mandelic acid and 0.003 mol% (*S*)-mandelic acid, respectively. In both droplets, the disclination line is curved. In case of Figure 5.3b the curvature is S-like, whereas the disclination line of the droplet in Figure 5.3c shows a 2 shape ("mirrored S"). A schematic director field for the achiral and both chiral cases is shown in Figure 5.4.

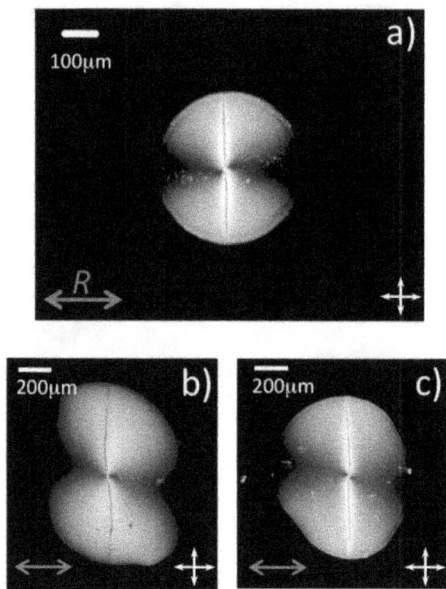

Figure 5.3: Polarized optical micrographs of Type I droplets of pure 5CB in (a), of 5CB doped with 0.018 mol% (*R*)-mandelic acid in (b); and of 5CB doped with 0.003 mol% (*S*)-mandelic acid in (c). The crossed polarizers are marked as white cross and the rubbing direction *R* is marked horizontal in red. Figure reprinted with permission from Langmuir from ref. [133].

Putting a sessile drop of an achiral nematic LC on a glass plate with a uniform planar alignment layer at the substrate surface, the director **n** of the LC orients parallel to the rubbing direction of the alignment layer. At the approximately spherical nematic LC-air interface the nematic director orients perpendicular (homeotropic), which makes the in-plane component of the director at this surface point radially away from the center, see Figure 5.4a.

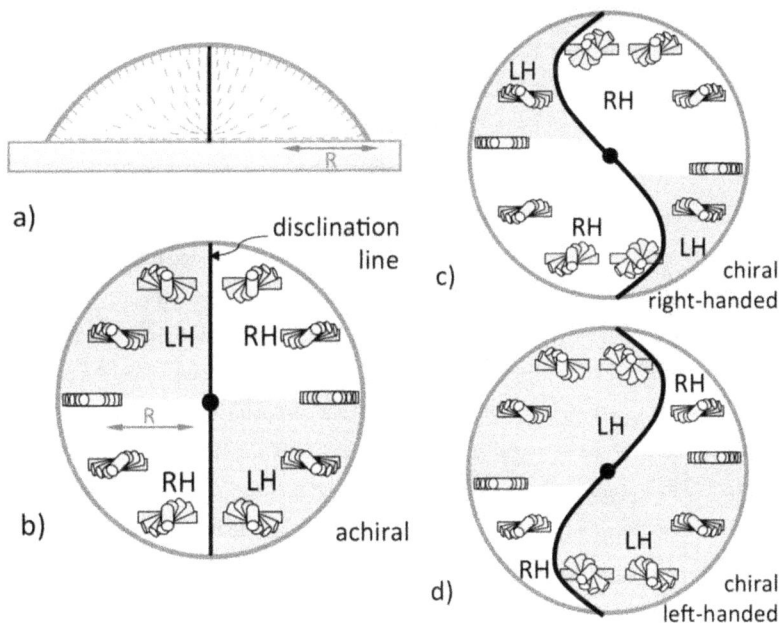

Figure 5.4: Schematic illustration of the director field within a sessile type I nematic LC droplet with a +1 disclination at the center marked as a black dot, in the achiral (a) – (b) and chiral (c) – (d) case. (a) Cross-section along the 9 – 3 diameter parallel to the rubbing direction *R*. In this cross-section plane the nematic director field only contains splay and bend deformations, but no twist deformation. (b) Top view of the achiral droplet illustrating the local hybrid twisted director field inside the drop. Normal to the rubbing direction a disclination line occurs across which the twist handedness changes. Left-handed (LH) and right-handed (RH) twist domains are indicated with grey and white respectively. The sense of twist changes discontinuously when moving parallel to the rubbing direction, but continuously when moving perpendicular to the rubbing direction. (c) and (d) show that the disclination line is deformed in the case of a chiral nematic LC. (c) An S-shaped disclination line reveals a right-handed twist. (d) A 2-like disclination line reveals a left-handed helical superstructure of the N* phase. Figure reprinted with permission from Langmuir from ref. [133].

133

The given boundary conditions of the droplet confinement – the planar anchoring at the LC-glass interface and the homeotropic anchoring at the LC-air interface – give rise to a hybrid splay-bend director field along the 9 – 3 diameter (clock-face notation) parallel to the rubbing direction. Along all other diameters, this hybrid structure is, additionally, twisted along the substrate normal, see Figure 5.4c,d. The twist, and therefore also the elastic energy density, is a function of the azimuthal angle and in order to minimize the elastic energy, the handedness of the twist transforms discontinuously from left- to right-handed (-90° to +90°) across a disclination line perpendicular to the rubbing direction. In the following type I and type II droplets will be distinguished by the presence of an additional +1 disclination at the center of the droplet, which is indicated as a black dot in Figure 5.4.

Chirally doped nematic liquid crystal droplets are shown in Figure 5.3b,c and in Figure 5.5. In comparison to achiral sessile droplets, the disclination line is curved and rotated relative to the rubbing direction about an angle α. The shape of the disclination line – either S- or 2-like – is directly related to the chirally induced helix sense in the N* phase. Comparing the droplets shown in Figure 5.3b,c and the ones in Figure 5.5, they can be distinguished by the presence of a central +1 disclination. The droplets of lower dopant concentration exhibit an additional +1 defect at the center of the drop and are referred to as type I droplets in the following. On the contrary, at higher dopant concentrations, this configuration seems to be unstable and the +1 disclination disappears and solely the curved disclination line remains. This configuration is called type II. The type II configuration is related to spherical nematic drops with homeotropic anchoring and an internal axial director field, characterized by a disclination loop along the equator.[59,164,165] The type II sessile droplet is obtained from such a spherical droplet by cutting it normal to the equatorial plane.

The director field within type I droplets is explained in Figure 5.4, whereas those of type II droplets are shown in Figure 5.6. When the nematic liquid host phase is chirally doped, the minimum in elastic energy of sessile droplets corresponds to a finite twist of the hybrid structure. Analogous to the experiment by Suh et al.[159], see Figure 5.2c, the disclination, which forms at the position of maximum elastic energy and is azimuthally positioned in the middle between regions of minimum elastic energy, is rotated.

In view of the fact, that the droplet height decreases when approaching the periphery, the twist and therefore the elastic energy density as well are simultaneously a function of the radial distance from the center. At the periphery, the thickness is small and the twist large; the disclination will experience a much less distortion than in the center. This gives a curved disclination line and the shape adopts either an S-form for a right-handed induced helix and a 2-form for a left-handed induced chiral nematic phase.

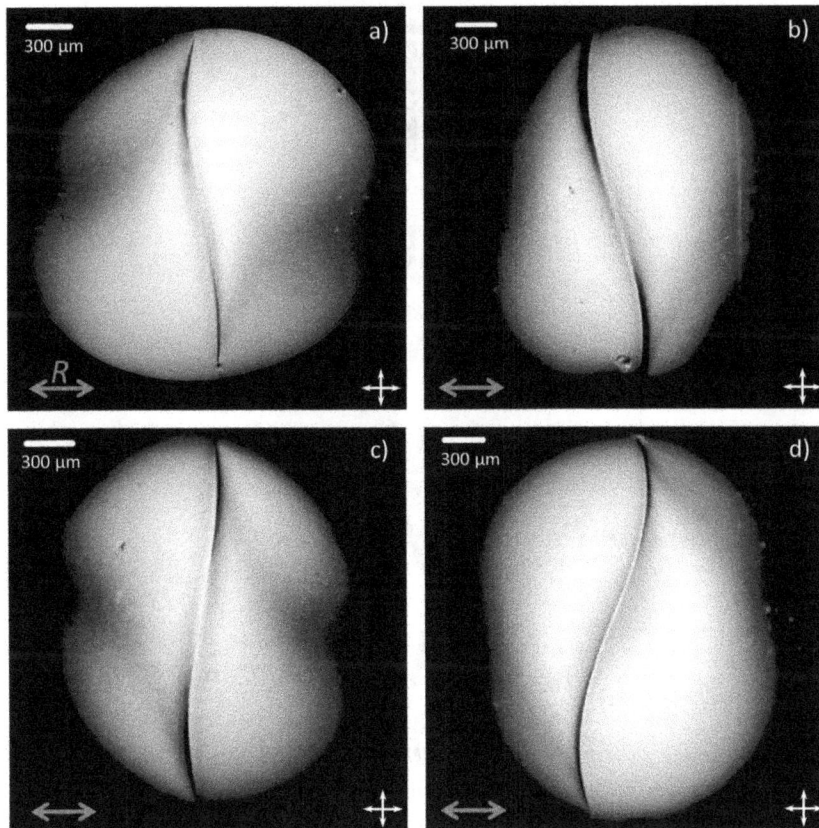

Figure 5.5: Polarized optical micrographs of type II sessile droplets of nematic liquid crystal 5CB doped with (R)-mandelic acid in (a) and (b) and with (S)-mandelic acid in (c) and (d). Comparing the disclination line shapes to the schematic director configurations shown in Figure 5.6, it can be directly seen that a right-handed chiral nematic phase has been induced by the addition of (R)-mandelic acid, whereas a left-handed chiral nematic phase has been induced by (S)-mandelic acid. The crossed polarizers are marked as a white cross and the rubbing direction R of the anisotropic polyimide coated glass surface is marked in red. Figure reprinted with permission from Langmuir from ref. [133].

135

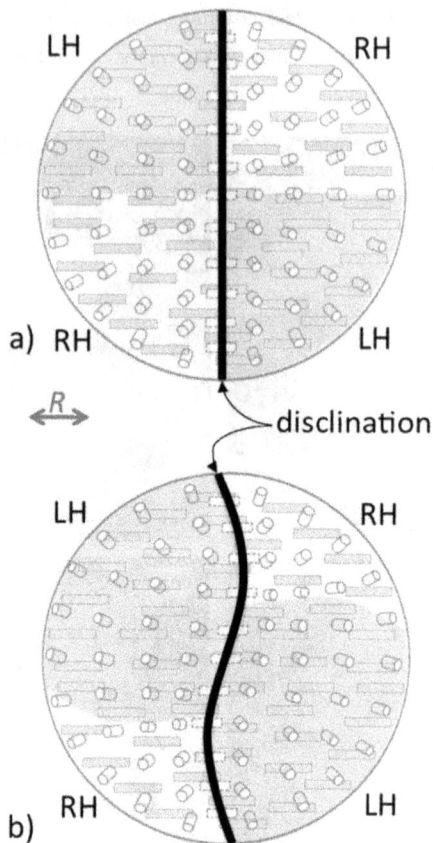

Figure 5.6: Top view illustration of the director field within a sessile type II nematic LC droplet with no +1 point defect in the center compared to type I droplets. (a) Schematic picture of an achiral nematic type II LC droplet. The rubbing direction *R* is marked in red. The mesogens at the flat glass surface are blue shaded, whereas the mesogens at the homeotropic spherical LC-air interface are white. The disclination line runs at the top of the surface of the droplet and the directly underlying mesogens are marked with dashed contours. (b) Schematic picture of a chiral nematic type II LC droplet in which the domains with the energetically favored twist sense induced by the chiral dopant enlarge at the expense of the domains with the opposite twist sense. The picture shows a left-handed induced twist of the director which gives a 2-form of the disclination line. Note that the expansion of the left-handed regions also occurs parallel to the rubbing direction. This is allowed due to a weak effective in-plane anchoring of the director at the homeotropic top surface. Figure reprinted with permission from Langmuir from ref. [133].

5.3.2 Pitch measurements

Apart from directly determining the handedness of the induced pitch from the shape of the disclination line, the magnitude of the pitch can be estimated as well by measuring the rotation angle α. Figure 5.7 shows how the rotation angle α is determined. An idealized S- or Ƨ-shaped disclination line contains a center of symmetry, namely an inflection point, in which the curvature of the disclination line changes sign. The angle α is the angle between the disclination line and the normal to the rubbing direction measured at the inflection point, see Figure 5.7a. In reality, the observed disclination is a little bit distorted from its idealized centrosymmetric shape; see Figure 5.7b. Here, the reference line must be drawn normal to the precisely known rubbing direction in the way that it intersects the disclination line in its inflection point. At the intersection point, the tangent to the disclination line is delineated and the angle α is measured between the reference line and the tangential line. In type I droplets, the intersection point coincides with the +1 defect, whereas in type II droplets it is found that the disclination line is more or less linear in the center of the droplet and the inflection point is situated in the middle of this linear regime. In a first approximation, the pitch of the chiral nematic LC is calculated as:

$$P = \frac{360°}{\alpha} h \ . \tag{73}$$

The error in the determination of the rotation angle α is about $\pm 0.5°$ and h is the droplet height which can be exactly measured by, e.g., a contact angle measurement setup, as it was done in Figure 5.7c.

However, in contrary to CR- and TN-cells in which pure twist deformation of the director field occurs, the more complex hybrid director field within a sessile droplet has to be considered. The *Helical Twisting Power HTP* [4] is valid or fully effective solely if the twist runs normal to the director. Along with the director, there can be no twist neither sustained nor induced. Therefore, in both hybrid arrangements of type I and type II droplets the effective *Helical Twisting Power* gradually decreases from *HTP* at the bottom to zero at the top of the droplets under the assumption of a homeotropic orientation of the director at the curved LC-air interface.

---

[4] To avoid misapprehension, the *Helical Twisting Power* is abbreviated as *HTP* in the continous text and as *H* in equations.

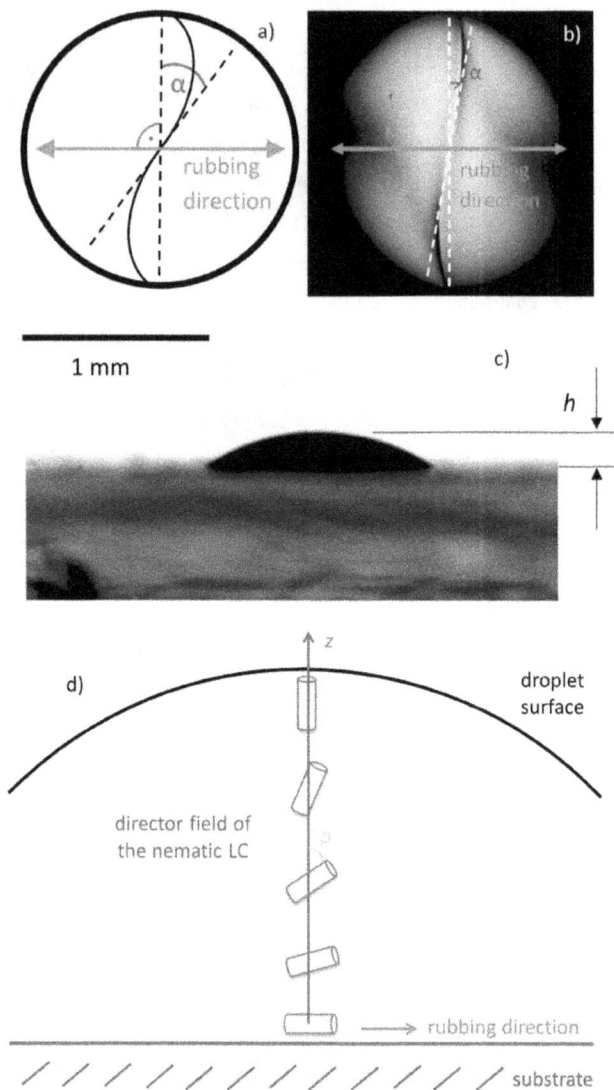

Figure 5.7: (a) and (b) Schematic illustration of measuring the rotation angle α of the disclination line. (c) Height *h* of the sessile droplet. The image was taken by means of a contact angle measurement setup. (d) Scheme of the local director field **n**(*z*) along the direction normal to the substrate *z* with the polar angle *ρ*. Sketch (a) – (c) is reprinted with permission from Langmuir from ref. [133].

A qualified guess of the effective *Helical Twisting Power* HTP_{eff} (abbreviation in equations: H_{eff}) along the normal to the flat glass substrate is therefore:[166]

$$H_{eff} = H \sin^2(\rho) \ . \tag{74}$$

with ρ as the polar angle between the local nematic $n(z)$ director and the direction normal to the substrate z, see Figure 5.7d. Similar to the situation in a hybrid aligned nematic cell, the splay-bend deformation adopts in the one-constant approximation a minimum elastic energy state in which ρ changes linearly ($\partial^2\rho/\partial z^2 = 0$) with height z between the planar and homeotropic anchoring at the LC-glass and LC-air interface respectively.[167] Under the assumption of a linear variation in ρ from 90° at the bottom to zero at the top of the droplet as a reasonable first approximation and average over the droplet height, HTP_{eff} can be expressed as:

$$H_{eff} \approx \frac{1}{2}H \ . \tag{75}$$

The effective *Helical Twisting Power* along the surface normal should be half of the value which is valid for the case when all twist goes normal to the director. This result modifies Equation (73) according to:

$$P \approx \frac{360° \ h}{\alpha \ 2} \ . \tag{76}$$

To avoid ambiguities in the determination of α, the total twist across the droplet should be small, in other words, the height h of the droplet should be smaller than the pitch which one would like to measure. Because the height of the droplet is typically around $100 - 200 \ \mu m$, as measured by a contact angle measurement setup and shown in Figure 5.7c, the sessile droplet method on anisotropic surfaces is suitable to measure very long pitch lengths – in the order of millimetres – in case of either very low dopant concentrations or mixtures of nematic LCs with dopants of very weak *Helical Twisting Power*.

By measuring the pitch of the induced chiral nematic liquid crystalline phase of mixtures with different dopant concentrations, the *Helical Twisting Power* of the dopant in this host material can be determined, according to Equation (10). Figure 5.8 shows sessile drops of 5CB with increasing (S)-mandelic acid concentration. The ones in Figure 5.8a,b are of mixtures of low dopant concentration and therefore appear in the type I configuration, whereas those in Figure 5.8c,d are of mixtures of higher dopant concentration and therefore appear in the type II configuration. However, the type of droplet does not affect the analysis of the pitch.

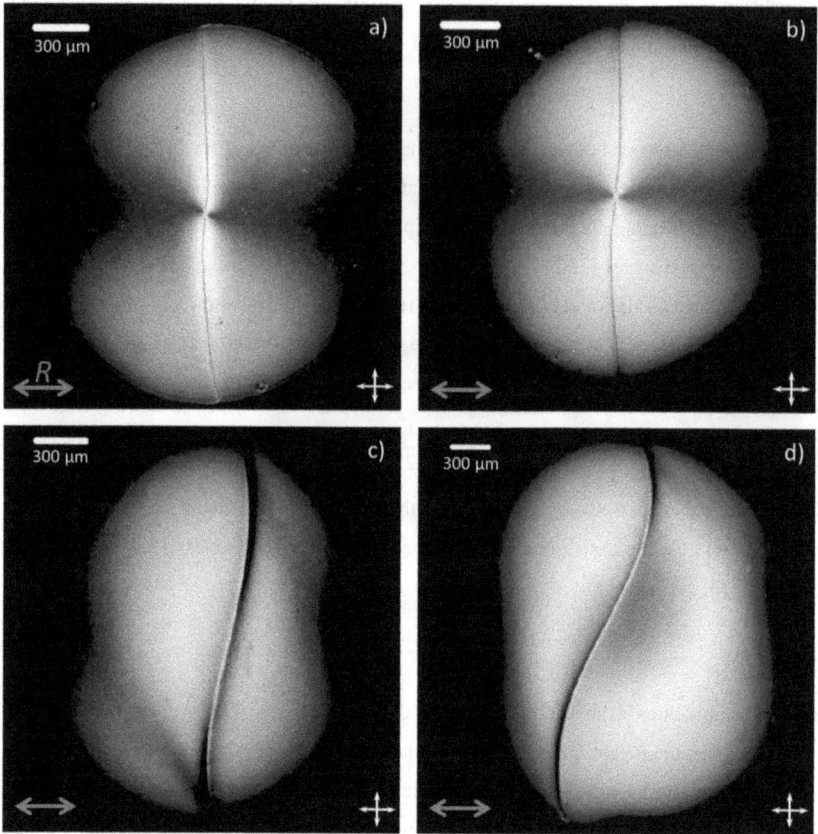

Figure 5.8: Polarized optical micrographs of sessile droplets of the nematic liquid crystal 5CB doped with (a) 5.44 mmol%; (b) 0.011 mol%; (c) 0.027 mol% and (d) 0.054 mol% of (S)-mandelic acid. In (a) and (b) the sessile drops are of type I configuration with a +1 disclination at the center, whereas in (c) and (d) the drops are of type II. The 2-like shape of the disclination line reveals a left-handed chiral nematic phase induced by (S)-mandelic acid. The crossed polarizers are marked as white cross and the rubbing direction R of the polyimide coated glass surface is marked in red. Figure reprinted with permission from Langmuir from ref. [133].

The pitch as a function of the (S)-mandelic acid concentration was determined by the rotation angle α and Equation (76) with data of both – type I and type II – droplets. For reference, the pitch was also measured with the TN cell method, see Figure 5.9. In Figure 5.9a a TN cell filled with achiral 5CB is shown, here the disclination lines pinned by spacer particles are straight. Figure 5.9b depicts a TN cell filled with a mixture of 5CB and 0.016 mol% of (*S*)-mandelic acid, in that case, the disclination lines are curved and the pitch is obtained from the curvature radius, see Equation (72). The results of both methods are shown in Figure 5.10. The red circles represent the pitch data obtained by the droplet method whereas the black squares represent the data obtained by the TN cell method. For pitch measurements via the droplet method, drops with approximately the same diameter were selected and analyzed, since these should have approximately the same height *h*. For one drop the height was exemplarily measured, see Figure 5.7c. The pitch (filled symbols) and the inverse pitch (open symbols) are plotted versus the mole fraction of (*S*)-mandelic acid as chiral dopant in Figure 5.10 giving a hyperbolic and linear regression respectively. The slope of the linear plot is proportional to the *Helical Twisting Power* of the dopant, see Equation (10). The values from the simple droplet method are in good agreement with the ones obtained with TN cells, although a little more scattered as the accuracy is smaller.

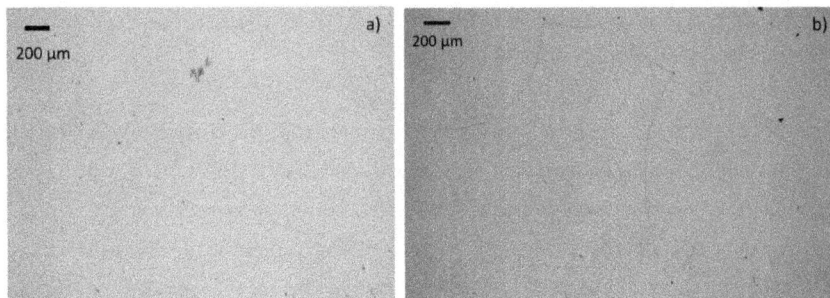

Figure 5.9: (a) TN cell filled with achiral 5CB. The disclination lines are pinning straight between spacer particles. (b) TN cell filled with a mixture of 5CB and 0.016 mol% of (*S*)-mandelic acid, in that case, the disclination lines are curved and the pitch can be obtained from the curvature radius. The curvature radius is measured with the image editing software GIMP by fitting the curved disclination line – which represents the arc of a circle – to circles with definite radii.

141

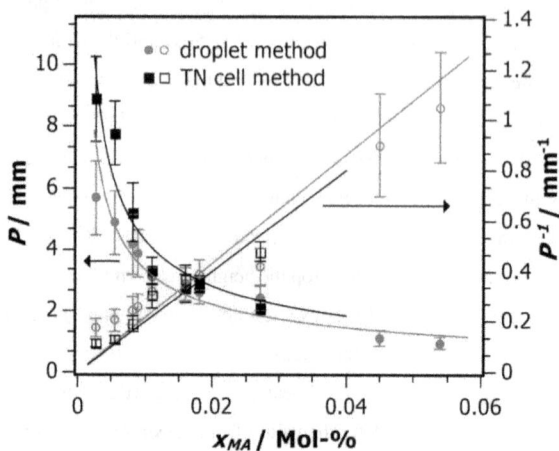

Figure 5.10: Pitch (filled symbols) and inverse pitch (open symbols) as a function of the molar fraction of (*S*)-mandelic acid in 5CB measured with the droplet method (red circles) and the TN cell method (black squares) as reference. The error of the pitch was calculated via standard deviation of the mean. Figure reprinted with permission from Langmuir from ref. [133].

Since for the pitch measurements the height h of the droplet should be smaller than the pitch, the minimum, as well as the maximum detectable rotation angle α, can be tuned up to a certain degree by increasing or decreasing the droplet height. Let us discuss the maximum detectable rotation angle α for droplets of $h \approx 100 - 200$ µm. In the experiments with 5CB as host phase and mandelic acid as chiral dopant, the droplet method was pushed to its upper limits at around 0.09 mol% of mandelic acid, see Figure 5.11. In case of too high dopant concentrations – and/or dopants of too high *Helical Twisting Power* – the disclination line shows a strong distortion up to a point where the inflection point cannot be precisely determined anymore and the hybrid twisted director field as discussed in Figure 5.4 and Figure 5.6 transforms into a more complex configuration in which half of the pitch is smaller than the droplet height, see especially Figure 5.11c. The droplet height increases from Figure 5.11a to Figure 5.11c.

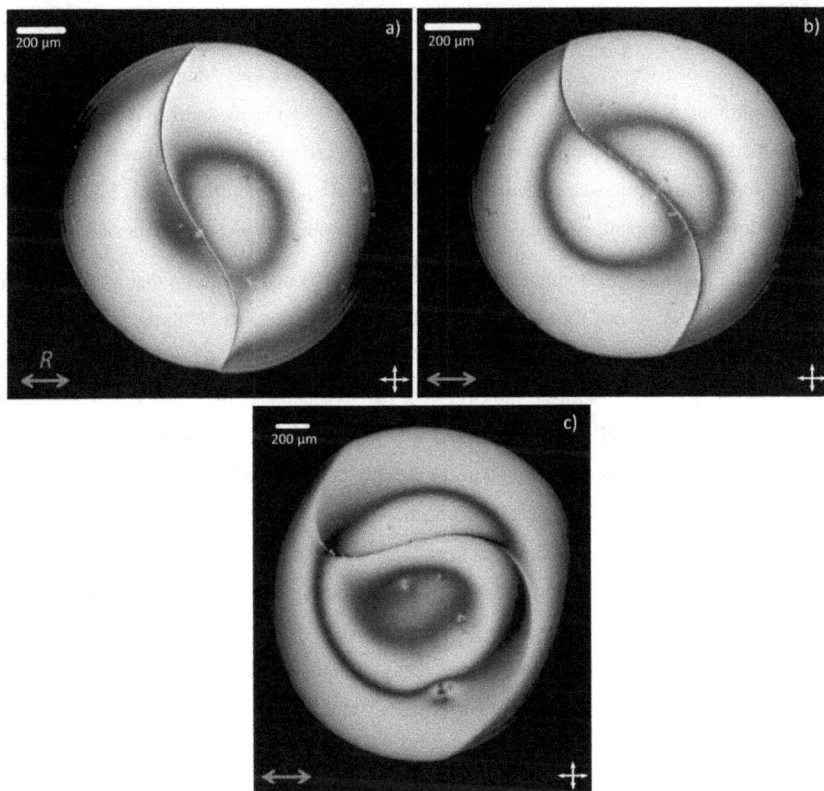

Figure 5.11: (a) – (c) Polarized optical micrographs of sessile droplets of the nematic liquid crystal 5CB doped with 0.09 mol% of (R)-mandelic acid. The droplet height increases from (a) to (c). The distortion of the disclination line increases with increasing droplet size and measurements of the rotation angle α become more and more inaccurate since the inflection point of the disclination line can not be precisely identified. The crossed polarizers are marked as white cross and the rubbing direction R of the anisotropic polyimide coated glass surface is marked in red.

5.4 Chapter Conclusion

The sessile droplet method opens new possibilities to measure very long pitch lengths up to several millimeters and to directly determine the induced twist handedness in chiral nematics. The method is strikingly simple. It neither requires advanced sample preparations nor advanced analytic tools. Furthermore, it consumes very little material, both of the nematic liquid crystal host and the added dopant, since the volume of typical droplets is just a few tens of nanoliters.

Investigating a mixture of 3 mmol% mandelic acid in 5CB and assuming that a drop with a volume of 20 nL is needed for analysis, the calculation is as follows: With the molar mass of 249.35 g/mol and density of 1.008 g/mL for 5CB this gives 8.08 x 10^{-8} moles (= 4.87 x 10^{16} molecules) of 5CB and 2.44 x 10^{-11} moles (1.46 x 10^{13} molecules) of mandelic acid. This means that for each drop only 20.1 µg of 5CB and 3.69 ng of mandelic acid is needed. The type I droplet configuration is directly related to the radially photo-buffed cells from ref. [160]. But, in the sessile droplets the radial in-plane component of the anchoring is due to the curved surface with homeotropic boundary conditions at the LC-air interface. Therefore, the effective in-plane anchoring at the top surface can be regarded as weak, going towards zero at the poles of the droplet. This allows the regions of energetically favored twist sense to expand at the expense of the energetically unflavoured twist sense areas, as indicated by shaded regions in Figure 5.4c,d, and Figure 5.6b. In the radially buffed cells[160] and also in the CR cells[159], the in-plane anchoring is strong and no change in the director field apart from the moving disclination line in comparison to the achiral case occurs.

Putting a large number of sessile droplets on a rubbed anisotropic glass surface, not all droplets show clean type I or type II structures, as it has been shown in Figure 5.5 and Figure 5.8. For example, dust particles can pin the disclination lines and prevent it to adopt S- or Ƨ-like shapes. In some droplets, the footprint is not circular, not even smooth in shape, which can be due to slipping of the pipette tip during the deposition process. In practice, when depositing the droplets by hand, it is found that around one half to one-third of the drops can be regarded as clean and useful for analysis. Some other complications can arise due to the wetting of the droplets on the glass substrate, i. e. a spreading along the rubbing direction.

It was found, that the wetting on glass plates with an underlying indium-tin-oxide (ITO) coating was much more problematic than on glass substrates where the ITO was removed by etching. The wetting can be attributed to a change in the surface anchoring conditions. The nematic director field reacts very sensitively to such small changes which can have drastic effects.

For example, changing the composition of the nematic liquid crystal by adding a chiral dopant can already change the elastic properties of the liquid crystal and affect the delicate interplay between the surface and the nematic director field.[117] However, importantly, S- and Ɛ-shaped disclination lines never occurred at the same time for a certain dopant handedness. And, both type I and type II drops showed always the same shape of disclination line. Therefore, it can be said that the droplet method is unambiguous with respect to the handedness of the induced chiral nematic phase when analyzing both type I and type II drops. Furthermore, both types give rather correct pitch values compared to the results obtained by TN cells, see Figure 5.10.

Due to the fact that it is a disclination line which is to be analyzed permits the use of an optical microscope even without polarizers. No moving parts are needed, e.g. neither rotation of the droplets nor any polarizer are necessary for measurements of the induced *Helical Twisting Power*. This opens the possibility for high throughput analysis by computational image analysis. Furthermore, by automatized image analysis, statistics over a larger amount of droplets could easily be done. Enhanced reproducibility of the deposition process, such as creating a uniform distribution of volumes and heights, can be achieved by using, e.g., an automatic glue dispenser or even an inkjet printer. Figure 5.12 shows an image of a set of sessile droplets deposited by an automatic glue dispenser. Only the lower left droplet is a bit larger than the rest, but overall the size-, and therefore height- distribution is quite narrow.

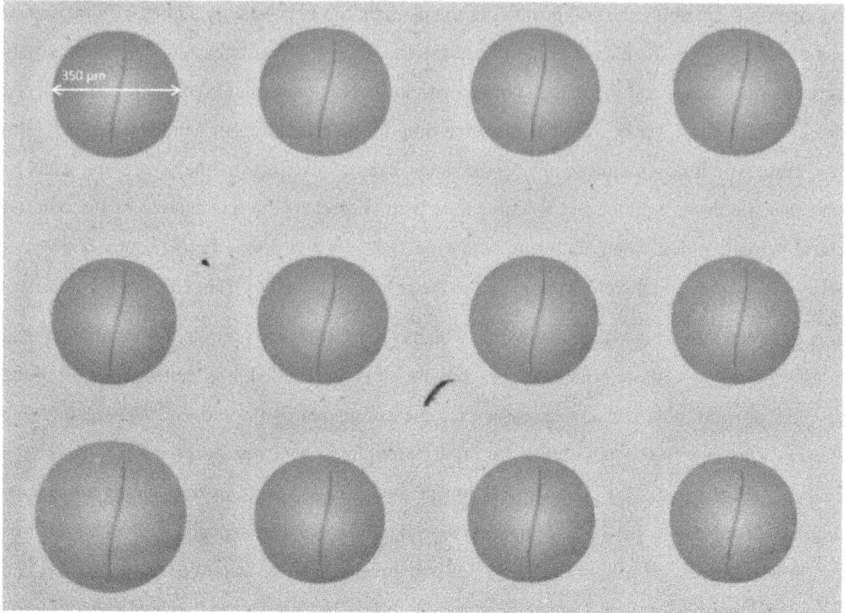

Figure 5.12: Micrograph of a set of sessile chiral nematic liquid crystalline droplets deposited by an automatic glue dispenser in the MC2 cleanroom facility of the Chalmers University of Technology. Photo by courtesy of Prof. Per Rudquist.

Comparing the TN cell method and the droplet method, it is evident that the optimized conditions for both methods are requiring a surface anchoring with exactly zero pretilt. Such boundary conditions can be realized by oblique evaporation of SiO_x or SiO_2 layers. For example, in ref. [163], a 60° oblique evaporated SiO film gave a disclination line that appears across the center of the sessile droplet.

In the droplet method, the accuracy in the pitch determination is limited by the errors in the measurements of the rotation angle $\alpha \pm 0.5°$, the exact droplet height h, which could be measured for each drop separately by a contact angle measurement setup, and the approximation of the effective *Helical Twisting Power* in the hybrid director structure in the drops. In clean type I and II droplets with a circular footprint and well-centered disclination lines, rotation angles of $\alpha \approx 2°$ can readily be measured. According to equation (65), the method then allows a pitch value estimation of about 20 mm in typical droplets of a height of 200 μm.

A major advantage of the droplet method is that the substrate is achiral. Hence, any observed chirality effects should directly originate from the studied material.

An interesting further development of the method could be to perform the chiral doping in ready-made sessile droplets and monitor the chiral induction process. One example could be that achiral sessile droplets are exposed to chiral vapors dissolving into the nematic liquid crystal. The evolution of the increasing chiral content in the droplets can be monitored by the distortion of the disclination line from straight to increasingly S- or ƨ-shaped. The shapes of the disclination lines of droplets at different locations could directly indicate and measure both the temporal and spatial distribution of exposure. Furthermore, the spread of gas by adding chiral molecules with very high *Helical Twisting Power* to an achiral gas can be measured and monitored by means of sessile droplet chirality sensors.

The doping of ready-made nematic LC drops might also be obtained by using the dopant in its liquid or solid state. The latter would be related to the work by Gvozdovsky et al.[168] who studied the addition of steroid crystals to large nematic droplets on a surface. Dissolving the steroid crystals into the nematic host, a rotation of the droplet structure with increasing chiral content can be directly observed. Another idea would be that the chiral doping could originate from the solid surface itself, e.g. revealing leakage or diffusion of chiral species from, or even through, a particular substrate.

Within the scope of the bachelor project by Nadine Schnabel,[169] the photoisomerization process and chirality effects in azobenzene doped nematic liquid crystals under confined geometries were studied via the CR cell method and the use of sessile droplets. Under the illumination of UV light, the rod-like and planar trans isomer of an azobenzene molecule transforms into the bent-shaped and non-planar cis isomer. The angular geometry of the cis isomer leads to the existence of axial helicity in the (*P*)- and (*M*)-enantiomer, see Figure 5.13.

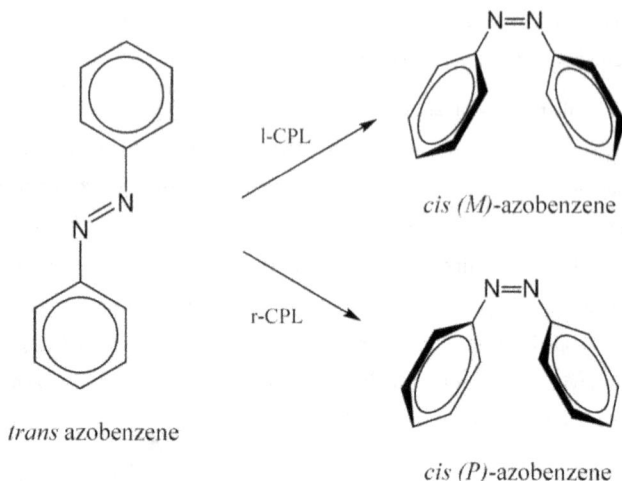

Figure 5.13: Mechanism of the enrichment of the (*P*)- or (*M*)-enantiomer of the cis azobenzene, by irradiation of either right-handed (r-CPL) or left-handed (l-CPL) circularly polarized UV light during photoisomerization. Figure reprinted with permission from ref. [169].

Among others, the idea was to investigate the possibility to induce chirality on a macroscopic scale in nematic liquid crystals by illuminating LC mixtures of 5CB doped with non-chiral azobenzenes with circularly polarized UV light of one handedness, functioning as external chiral action, which should lead to an enantiomeric excess of either the (*M*)- or (*P*)-enantiomer of the azobenzene cis isomer.

For this investigation, the droplet method was chosen because of its simplicity and sensitivity to detect an even very small enrichment of one of the enantiomers. Even though the effect observed for a single drop is a bit too ambitious, a statistical approach using an automatic glue dispenser to assemble a large number of drops of defined height, size, and shape, as shown in Figure 5.12, was leading to some first experimental result. Before UV illumination, the disclination line is straight, as expected because the LC, as well as the trans isomer of the azobenzene, are achiral. However, after illuminating the samples with left-handed circularly polarized UV light, the number of droplets with straight disclination lines decreased strongly whereas the number of, e.g., S-shaped disclination lines increased from zero to over 50%.

On the contrary, by illuminating the samples with right-handed circularly polarized UV light the number of drops with the reversed shaped disclination line increased over 50%. This means, that the irradiation with circularly polarized UV light induces a change in the disclination shape which is directly related to the handedness of a possibly induced chiral nematic phase. Those results can be regarded as a first indication that a small enantiomeric excess of either the (*P*)- or (*M*)-enantiomer of the cis isomer was induced by circularly polarized light of the respective handedness and that this chirality effect happening on the microscopic scale has been amplified to a macroscopic scale by means of a nematic liquid crystalline host which forms a helical superstructure in the presence of chiral molecules. However, because these results are based on just a statistical approach, further investigations like e.g. examining the electroclinic effect in a SmA phase could give more detailed and clear evidence.[169]

In conclusion, the sessile droplet method is a highly sensitive and simple method for the detection of weak chiral induction in thermotropic nematic LC hosts. The sensitivity in determining the sign of the induced twist sense is based on the easiness of simple distinction between an S- and 2-shaped disclination line. The method requires for a single droplet only a sub microliter volume of material and any substrates with conventional rubbed polymer, or obliquely evaporated inorganic, alignment coating can be used.

6 Summary and Outlook

In this thesis, chirality effects in thermotropic and lyotropic liquid crystals under confined geometries are studied. It is shown that by means of the boundary conditions of the surrounding confinement, director configurations of nematic liquid crystals can be obtained which are extremely sensitive towards the addition of chiral molecules and thus to chiral induction. The process of chiral induction describes the chirality transfer from the microscopic scale of the chiral dopant to the macroscopic scale of the induced chiral liquid crystalline phase; this process is still not fully understood.[57,58,147,148] The director configurations obtained by using certain confinements are applied to measure qualitatively and quantitatively the characteristics of a chiral nematic phase such as the handedness of the induced helical superstructure and its pitch. Therefore, the use of those configurations provides further insight into the still unclear process of chiral induction and offers new possibilities to detect weak chirality effects at very small dopant concentrations, for which, e.g., only one out of 3000 molecules is chiral.

At first, a standard micellar lyotropic LC was subjected to capillary confinement with homeotropic boundary conditions, reflecting the idea that the pitch of a possible helical superstructure can develop unhindered over the range of several centimeters along the long axis of a capillary. Interestingly, it was found that the achiral LLC is forming chiral director configurations without the addition of a chiral dopant. This occurrence of chiral structures in a system containing solely achiral components is known as spontaneous mirror symmetry breaking and is of special interest among all disciplines of natural science because it is related to the origin of homochirality in nature which is still an unsolved issue.

Recently, in the field of LCs, spontaneous reflection symmetry broken configurations in achiral chromonic LCs under capillary confinement were reported and their appearance was attributed to a small twist elastic modulus which was found to be one order of magnitude smaller than the splay and bend moduli.[61,64,68] In the case of the standard micellar lyotropic LC investigated in this thesis, similar chiral configurations were observed,[45] which suggests that a similar elastic anomaly in standard micellar LLCs exists.

One of these chiral configurations is the twisted escaped radial (TER) configuration which is the chiral analog to the well-known escaped radial configuration. Under the action of a magnetic field, another chiral configuration is generated, which is called twisted polar (TP) configuration. It is characterized by two half-unit twist disclinations for which the director twist around the line defects drives the formation of a double helix of the disclinations along the long axis of the capillary.[45]

Since the LLC is still achiral, both left- and right-handed helices occur with the same probability. However, by adding a chiral dopant, e.g., (R)-mandelic acid, the energetic degeneracy between left- and right-handed twist senses is lifted and the delicate balance is biased towards one handedness. This evolves into an extreme sensitivity for chirality in those configurations, which can be exploited. In the exemplary case of using (R)-mandelic acid as a dopant, different regimes of chiral induction are revealed. In the concentration range of 6 mmol% – 90 mmol% of (R)-mandelic acid the TER configuration develops from a homochiral TER to a doubly twisted escaped radial configuration and finally to the well-known fingerprint texture.

In the case of the magnetic field-induced twisted polar configuration, the following four regimes of chiral induction have been found: (i) At very low dopant concentration there is a heterochiral regime. Both twist senses occur with the same probability despite the addition of a chiral dopant. The pitch of the double-helical twisted polar configuration is only marginally influenced by the dopant, meaning that it is mainly set by the capillary dimensions and the ratios of elastic constants. (ii) Upon a further increase of (R)-mandelic acid concentration, the handedness of the TP configuration is influenced by the handedness of the chiral dopant, biasing the twist sense of the TP on the scale of centimeters. (iii) Then the pitch of the TP is influenced by the *Helical Twisting Power* of the chiral dopant, which causes a drastic decrease in its periodicity. (iv) Furthermore, the well-known fingerprint texture appears in addition, taking over the intrinsic influence of the dopant on the LLC.

In conclusion, both chiral configurations, i.e., the TER as well as the TP configuration, constitute a very sensitive tool to investigate chirality effects like the chiral induction from a chiral dopant to an achiral host phase in nematic lyotropic LCs. Like in chromonic LLCs, this is on one hand due to confirmed elastic peculiarities of the lyotropic liquid crystal exhibiting a small twist elastic constant relative to the splay and bend constants which leads to the spontaneous formation of chiral structures already in the achiral case. On the other hand, the geometrical support of the capillary confinement permits the formation and propagation of a helix axis along the long axis of the capillary.

Furthermore, the viscoelastic properties of the micellar LLC were measured via depolarized dynamic light scattering, following the procedure of Zhou et al.[68]. The results render a splay constant K_1 of 1.79±0.7 pN, a twist constant K_2 of 0.33±0.01 pN and a bend constant K_3 of 20.8±1.0 pN. The relatively high bend elastic constant can be explained by its characteristic divergence upon approaching a lamellar phase.[118] The splay constant shows no critical behavior near a nematic-lamellar transition. More surprisingly, the twist elastic constant of the standard micellar LLC is one order of magnitude smaller than the other two elastic moduli. Compared to the known literature values of K_2 for lyotropic chromonic and polymeric LCs, this is to my knowledge the LC with the smallest twist elastic constant measured so far. Therefore the light scattering results confirm that the occurrence of director configurations with broken mirror symmetry can be explained by a small twist elastic constant – similar to the case of chromonics.

The system thus tends to release energetically costly splay and bend deformations into twisted ones by forming equilibrium helical structures. This phenomenon of a small twist elastic constant – and therefore the occurrence of mirror symmetry broken configurations – seems to be not restricted to the special case of chromonic LLCs. It seems to be a more general property of lyotropic LCs. However, the physics behind this remarkable difference between lyotropic and thermotropic LCs is still to be understood. Since the presence of a solvent makes the essential difference between lyotropic and thermotropic LCs, one might suspect that the solvent can serve as lubricant, as reported in the literature for a swollen SmC phase.[126] As discussed by Zhou et al.[56,68], the other significant difference between lyotropic and thermotropic LCs is the flexibility of the mesogenic building blocks and thus the much stronger fluctuations. The process of micelle formation is a continuous equilibrium in which surfactant molecules are continuously exchanged between a micelle and the surrounding solvent and then incorporated in neighboring micelles. Nonetheless, a conclusive explanation for this phenomenon is still to be found.

As a further future investigation, it would be interesting to clarify whether a standard micellar N_C phase and other nematic phases of standard micellar ionic or non-ionic systems, e.g., SDS/DOH/H2O, show the same anomaly of elastic constants and whether there is a dependence on the size of the building blocks. Furthermore, lyotropic systems with a high content of solvent and those with a low content of solvent could be compared with respect to their viscoelastic properties.

Concerning the director configurations of the LLC observed under capillary confinement, the formation process of the TP configuration still requires dedicated investigations. Additionally, it would be interesting to investigate the director configurations under capillary confinement for planar anchoring conditions. This could be easily carried out by treating the inner glass surface with a coating. This might lead to similar results like the ones reported in the case of chromonic LLCs.[64,66]

Since thermotropic LCs do not exhibit such an anomaly of the elastic moduli, the TER and TP configurations under capillary confinement cannot be observed in, e.g., the nematic 5CB liquid crystal. In contrast to lyotropic LCs, thermotropic LCs can be exposed to air without any concerns. This opens new possibilities concerning the use of other confined geometries involving the air-LC interface as an additional, independent boundary condition, like, e.g., the confinement of a sessile LC droplet on an anisotropic surface.

In the last part of this thesis, a powerful new method for the analysis of chiral induction in thermotropic nematics is introduced. This method makes use of small sessile nematic droplets and reveals the handedness and the pitch of a chirally induced helical twist in a single simple experiment. Most remarkably, the method is able to detect very long pitches up to 20 mm and thus allows to study the effects of very low dopant concentrations.[133]

In conclusion, this thesis shows that by means of the topology imposed by the confining geometry and by interfacial boundary conditions, unusual twisted director configurations in thermotropic and lyotropic nematic LCs are generated. Those director configurations can be exploited to investigate chirality effects. In this thesis, two new, very sensitive methods to study chiral induction for both classes of liquid crystals – for thermotropic as well as for lyotropic LCs – are presented. They are analyzed, inter alia, with respect to the types of geometrical confinements used, e.g., how the confinement amplifies, induces, and influences the detection of chirality effects.

7 Zusammenfassung und Ausblick

In dieser Dissertation wurden Chiralitätseffekte in thermotropen und lyotropen Flüssigkristallen in beschränkten Geometrien untersucht. Es wurde gezeigt, dass mithilfe von definierten Randbedingungen beschränkter Geometrien, Direktorkonfigurationen nematischer Flüssigkristalle erzeugt werden konnten, die sehr empfindlich auf die Zugabe chiraler Moleküle und chiraler Induktion reagieren. Der Vorgang der chiralen Induktion beschreibt die Übertragung der Chiralitätsinformation von der mikroskopischen Ebene des chiralen Dotierstoffs auf die makroskopische Ebene der induziert chiralen Phase eines Flüssigkristalls (LC). Dieser Vorgang ist bis heute noch nicht vollständig aufgeklärt und verstanden.[57,58,147,148]

Demnach kann durch Verwendung dieser Direktorkonfigurationen weitere Einsicht in den noch unklaren Prozess der chiralen Induktion erlangt werden. Zudem bieten diese Direktorkonfigurationen neue Möglichkeiten, um sehr schwache Chiralitätseffekte bei sehr kleinen Dotierstoffkonzentrationen, z. B. wenn nur eins von 3000 Molekülen chiral ist, zu beobachten und untersuchen.

Zunächst wurde ein mizellarer lyotroper Flüssigkristall in zylindrische Kapillaren mit homeotropen Randbedingungen gefüllt. Dabei war der Hintergedanken, dass sich der Pitch einer möglichen helikalen Überstruktur über die Länge von Zentimetern entlang der Kapillarachse ungestört ausbreiten kann. Jedoch wurde überraschenderweise beobachtet, dass der achirale lyotrope Flüssigkristall (LLC) bereits ohne Zugabe eines chiralen Dotierstoffs chirale Direktorkonfigurationen ausbildete. Dieses Phänomen, dass chirale Strukturen in einem System, das ausschließlich aus achiralen Komponenten besteht, vorkommen, ist bekannt als spontane Symmetriebrechung. Dies ist generell in der naturwissenschaftlichen Forschung von Relevanz und Interesse, da es mit dem Ursprung der Homochiralität in der Nature verknüpft ist, was bis heute ein noch ungeklärtes Thema ist.

Kürzlich, auf dem Gebiet der Flüssigkristallforschung, wurden chirale Konfigurationen in einer zylindrischen Kapillargeometrie für achirale chromonische Flüssigkristalle beobachtet. Deren Auftreten wurde anhand einer außergewöhnlich kleinen twist elastischen Konstante erklärt.

Während die splay and bend elastischen Konstanten in der gleichen Größenordnung liegen, ist die twist elastische Konstante in diesen chromonischen LLCs um eine Größenordnung kleiner. [61,64,68]

Im Falle eines gewöhnlichen mizellaren lyotropen Flüssigkristalls wurden in dieser Dissertation ähnliche chirale Direktorkonfigurationen entdeckt.[45] Dies legt den Schluss nahe, dass mizellare LLCs ebenfalls eine Anomalie der elastischen Eigenschaften aufweisen.

Eine dieser chiralen Direktorkonfigurationen ist die sogenannte twisted escaped radial (TER) Konfiguration. Sie stellt das chirale Analogon zu der bekannten escaped radial (ER) Konfiguration dar. Mithilfe eines Magnetfelds konnte eine weitere chirale Konfiguration, die sogenannte twisted polar (TP) Konfiguration, erzeugt werden. Sie zeichnet sich durch zwei halbzahlige twist Disklinationen aus. Hervorgerufen durch den Direktortwist um die Liniendefekte, bilden diese beiden Disklinationen eine Doppelhelix entlang der Kapillarachse aus. Da der LLC achiral ist, kommen in beiden Konfigurationen links- und rechtshändige Helixdrehsinne gleichhäufig vor. Jedoch kann durch Zugabe eines chiralen Dotierstoffs, wie z. B. (R)-Mandelsäure, die energetische Entartung dieser beiden Helixdrehsinne aufgehoben und hin zu einer Händigkeit beeinflusst werden. Dies führt zu einer extremen Chiralitätsempfindlichkeit dieser Konfigurationen, was sich am Beispiel von (R)-Mandelsäure als Dotierstoff zu nutzen gemacht werden kann. Es werden bei Zugabe von (R)-Mandelsäure vier verschiedene Regime der chiralen Induktion beobachtet. Im Falle der TER Konfiguration und eines Konzentrationsbereich von 6 mmol% – 90 mmol% an (R)-Mandelsäure wandelt sich die heterochirale TER in eine homochirale TER Konfiguration, geht in eine doubly twisted escaped radial (DTER) Konfiguration über, bis schließlich die bekannte Fingerprint Textur entsteht.

Im Falle der mithilfe eines Magnetfelds induzierten twisted polar (TP) Konfiguration, wurden folgende vier Bereiche der chiralen Induktion beobachtet: (i) Bei sehr kleinen Dotierstoffkonzentrationen ist die TP Konfiguration heterochiral. Das bedeutet, dass trotz Zugabe eines chiralen Dotierstoffs beide Helixdrehsinne mit der selber Wahrscheinlichkeit auftreten. Der Pitch der TP Konfiguration wird hauptsächlich durch die Kapillardimensionen und die Verhältnisse der elastischen Konstanten des LLCs bestimmt. (ii) Bei weiterer Erhöhung der Dotierstoffkonzentration wird der Helixdrehsinn der TP Konfiguration durch die Händigkeit des Dotierstoffs über mehrere Zentimeter hinweg bestimmt. (iii) Als nächstes wird der Pitch der TP Konfiguration durch die *Helical Twisting Power* des Dotierstoffs bestimmt, was zu einer drastischen Schrumpfung des Pitchs führt. (iv) Zuletzt tritt zusätzlich noch die Fingerprint Textur auf. Nun bestimmt der intrinsische Einfluss des Dotierstoffs die Periodizität der Fingerprint Textur des LLCs.

Zusammenfassend stellen beide chiralen Direktorkonfigurationen, die twisted escaped radial (TER) als auch die twisted polar (TP) Konfiguration, überaus empfindliche Hilfsmittel dar, um Chiralitätseffekte, wie z. B. die chirale Induktion eines chiralen Dotierstoffs auf die achirale Wirtsphase eines nematischen LLCs, zu untersuchen. Dies ist zum einen, wie bei den chromonischen LLCs, auf die in dieser Arbeit bestätigte Anomalie der elastischen Konstanten von lyotropen LCs zurückzuführen, was bereits im achiralen Fall zur spontanen Bildung helikaler Strukturen führt. Zum anderen erlaubt die zylindrische Kapillargeometrie die ungestörte Ausbildung einer Helixachse entlang der Kapillarachse.

Wie schon bereits kurz erwähnt, wurden in dieser Dissertation zusätzlich die viskoelastischen Eigenschaften des verwendeten mizellaren nematischen LLCs mittels depolarisierter dynamischer Lichtstreuung gemessen. Dabei wurde nach der Methode von Zhou et al.[68], in deren Arbeit die viskoelastischen Eigenschaften für chromonische LLCs bestimmt wurden, vorgegangen. Die Ergebnisse belaufen sich auf eine splay elastische Konstante K_1 von 1.79 ± 0.7 pN, eine twist elastische Konstante K_2 von 0.33 ± 0.01 pN und eine bend elastische Konstante K_3 von 20.8 ± 1.0 pN. Der verhätnismäßig hohe Wert für die bend elastische Konstante kann durch die Nähe einer lamellaren Phase erklärt werden. Während die splay elastische Konstante kein kritisches Verhalten in der Nähe eines nematisch-lamellaren Phasenübergangs zeigt, divergiert die bend elastische Konstante einer nematischen Phase bei Annäherung an eine lamellare Phase. Überraschenderweise ist die twist elastische Konstante eine Größenordnung kleiner als die anderen beiden elastischen Konstanten. Verglichen mit den literaturbekannten K_2 Werten für chromonische und poylmer nematische LLCs, ist dies der bisher geringste gemessene Wert einer twist elastischen Konstante. Demnach bestätigen die Ergebnisse der Lichtstreuung das Auftreten spontaner symmetriegebrochener Konfigurationen in diesem mizellaren LLC, analog zu dem Fall chromonischer LLCs.

Das lyotrope System tendiert dazu, energetisch ungünstige splay und bend Deformationen in twist Deformationen zu übertragen, indem es helikale Stukuren ausbildet. Dieses Phänomen einer kleinen twist elastischen Konstante – und demnach dem Auftreten spontaner symmetriegebrochener Konfigurationen – scheint nicht auf den speziellen Fall chromonischer

LLCs beschränkt zu sein, sondern eine generelle Eigenschaft lyotroper LCs zu sein. Nichtsdestotrotz ist die Physik hinter diesem außergewöhnlichen Unterschied zwischen thermotropen und lyotropen LCs noch nicht aufgeklärt und verstanden. Da das Vorhandensein eines Lösungsmittels der essentielle Unterschied zwischen lyotropen und thermotropen LCs darstellt, könnte man vermuten, dass das Lösungsmittel in LLCs als Schmiermittel fungiert, wie es in der Literatur am Beispiel einer gequollenen SmC-Phase diskutiert wurde.[126] Ein weiterer signifikanter Unterschied zwischen lyotropen und thermotropen LCs besteht in der Flexibilität der mesogenen Einheiten, siehe Diskussion von Zhou et al.[56,68]. Der Mizellbildungsprozess ist ein fortlaufendes Gleichgewicht, in dem die Tensidmoleküle kontinuierlich zwischen Mizelle und Lösungsmittel ausgetauscht, um dann in benachbarten Mizellen wieder eingebaut zu werden. Eine endgültige Erklärung für dieses Phänomen wurde jedoch noch nicht gefunden.

Es wäre interessant in weiterführenden Untersuchungen zu klären, ob eine mizellare N_C Phase oder andere nematische Phasen herkömmlicher mizellarer ionischer oder nicht-ionischer Systeme, wie z. B: das von SDS/DOH/H_2O, die gleiche Anomalie der elastischen Konstanten zeigen. Zudem könnte untersucht werden, ob es auch eine Abhängigkeit in der Größe der mesogenen Einheiten gibt, oder ob der Lösungsmittelgehalt des Systems eine Rolle spielt, indem man LLCs mit höherer und mit niedriger Lösungsmittelkonzentration miteinander vergleicht in Bezug auf ihre viskoelastischen Eigenschaften.

Was die Ausbildung der Direktorkonfigurationen des LLCs in zylindrischer Kapillargeometrie betrifft, wäre eine nähere Untersuchung des Bildungsprozesses der TP Konfiguration von Interesse. Zusätzlich könnte der LLC in Kapillaren mit planaren Randbedingungen gefüllt werden, was relativ einfach durch eine Oberflächenbeschichtung der inneren Glasoberfläche realisiert werden kann. Dies könnte zu ähnlichen chiralen Konfigurationen führen, wie es für chromonische LLCs beobachtet wurde.[64,66]

Da thermotrope LCs, wie z. B. 5CB, keine Anomalie der elastischen Konstanten aufweisen, können die TER und TP Konfigurationen in der Kapillargeometrie nicht beobachtet werden. Jedoch können thermotrope LCs – im Gegensatz zu lyotropen LCs – ohne Bedenken an Luft grenzen. Dies eröffnet neue Möglichkeiten bezüglich der Verwendung beschränkter Geometrien, in denen die Luft-LC Grenzfläche als zusätzliche Randbedingung in Betracht gezogen werden kann, wie z. B. im Falle eines sitzenden Tropfens auf einer anisotropen Oberfläche.

Im letzten Teil dieser Dissertation wurde eine überaus empfindliche Methode entwickelt, mit der die chirale Induktion in thermotropen LCs untersucht werden kann. Diese Methode basiert auf einem kleinen Tropfen auf einer anisotropen Glasoberfläche, anhand dem direkt die Händigkeit und der Pitch einer chiral induzierten nematischen Phase in einem einzigen, einfachen Experiment bestimmt werden können. Bemerkenswerterweise ist diese Methode dazu geeignet, sehr lange Pitches bis hin zu 20 mm detektieren zu können. Demnach ist sie prädestiniert, um Chiralitätseffekte bei sehr kleinen Dotierstoffkonzentrationen zu untersuchen. [133]

Zusammenfassend zeigt diese Dissertation, dass durch die Verwendung von verschiedenen beschränkten Geometrien und unterschiedlichen Randbedingungen an Grenzflächen, ungewöhnliche verdrillte Direktorkonfigurationen erzeugt werden können – sowohl in thermotropen als auch in lyotropen LCs. Diese Direktorkonfigurationen können sich zunutze gemacht werden, um Chiralitätseffekte zu untersuchen. In dieser Arbeit wurden zwei neue, überaus empfindliche Methoden entwickelt und dargestellt, mit denen die chirale Induktion in lyotropen und thermotropen LCs untersucht werden kann. Diese wurden hinsichtlich der Art der beschränkten Geometrie analysiert, z .B. wie die verwendete beschränkte Geometrie Chiralitätseffekte verstärkt, induziert und beeinflusst.

References

(1)   W. T. Kelvin. *Baltimore Lectures on Molecular Dynamics and the Wave Theory of Light*; CJ Clay and Sons: London, 1904.

(2)   F. Reinitzer. Beiträge zur Kenntniss des Cholesterins. *Monatsh. Chem.* **1888**, *9*, 421–441.

(3)   O. Lehmann. Über fliessende Krystalle. *Z. Phys. Chem.* **1889**, *4*, 462–472.

(4)   G. Heppke, C. Bahr. In *Bergmann, Schaefer, Lehrbuch der Experimentalphysik: Flüssigkristalle*; W. Raith, Ed.; Walter de Gruyter: Berlin/New York, 1992.

(5)   H.-D. Dörfler. *Grenzflächen und kolloid-disperse Systeme*; Springer Verlag: Berlin/Heidelberg, 2002.

(6)   H. Stegemeyer, Ed. *Lyotrope Flüssigkristalle: Grundlagen, Entwicklungen, Anwendungen*; Steinkopff-Verlag: Darmstadt, 1999.

(7)   F. Livolant, A. M. Levelut, J. Douct, J. P. Benoit. The highly concentrated liquid-crystalline phase of DNA is columnar hexagonal. *Nature* **1989**, *339*, 724–726.

(8)   M. Nakata, G. Zancheta, B. D. Chapmann, C. D. Jones, J. O. Cross, R. Pindak, T. Bellini, N. A. Clark. End-to-End Stacking and Liquid Crystal Condensation of 6-to 20-Base Pair DNA Duplexes. *Science* **2007**, *318*, 1276–1279.

(9)   G. Weerts. 3D gets a second look. *SPIE Professional* **2012**.

(10)   M. E. McConney, V. P. Tondiglia, J. M. Hurtubise, L. V. Natarajan, T. J. White, and T. J. Bunning. Thermally Induced, Multicolored Hyper-Reflective Cholesteric Liquid Crystals. *Adv. Mater.* **2011**, *23*, 1453–1457.

(11)   D. J. D. Davies, A. R. Vaccaro, S. M. Morris, N. Herzer, A. P. H. J. Schenning, and C. W. M. Bastiaansen. A Printable Optical Time-Temperature Integrator Based on Shape Memory in a Chiral Nematic Polymer Network. *Adv. Funct. Mater.* **2013**, *23*, 2723–2727.

(12)   S. T. Kim and H. Finkelmann. Cholesteric liquid single-crystal elastomers (LSCE) obtained by the anisotropic deswelling method. *Macromol. Rapid Commun.* **2001**, *22*, 429–433.

(13)   T. Ohzono, T. Yamamoto, J.-i. Fukura. A liquid crystalline chirality balance for vapours. *Nature Commun.* **2014**, *5*, 3735.

(14)   P. G. de Gennes, J. Prost. *The Physics of Liquid Crystals,* 2nd ed.; Clarendon Press: Oxford, 1993.

(15)   L. M. Blinov. *Structure and Properties of Liquid Crystals*; Springer Verlag: Heidelberg/London, 2011.

(16)   L. D. Landau. *Collected papers,* 1st ed.; Pergamon Press: Oxford, New York, 1965.

(17)   L. D. Landau. Theory of phase changes. I. *Zh. Eksp. Teor. Fiz.* **1937**, *7*, 19–32.

(18)   C. W. Oseen. The Theory of Liquid Crystals. *Trans. Faraday Soc.* **1933**, *29*, 883–899.

(19)   H. Zocher. *Trans. Faraday Soc.* **1933**, *29*, 945.

(20)   F. C. Frank. Liquid Crystals: Theory of Liquid Crystals. *Discuss. Faraday Soc.* **1958**, *25*, 19–28.

(21)   I. Dierking. *Textures of Liquid Crystals*; Wiley-Vch: Weinheim, 2003.

(22)   M. Reichenstein. *Dynamik von Disklinationen in anisotropen Fluiden.* Dissertation: Stuttgart, 2002.

(23)   A. D. Rey, M. M. Denn. Dynamic Phenomena in Liquid-Crystalline Materials. *Annu. Rev. Fluid Mech.* **2002**, *34*, 233–266.

(24)   C. Dietrich. *Chirale Induktion lyotrop-cholesterischer Phasen durch helikale Nanopartikel.* Masterarbeit: Stuttgart, 2014.

(25)   H. de Vries. Rotatory power and other optical properties of certain liquid crystals. *Acta Crystallographica* **1951**, *4*, 219–226.

(26)   J. Bruckner. *Struktur und Chiralitätseffekte in lyotrop-flüssigkristallinen Phasen eines 1,2-Diols*. Diplomarbeit: Stuttgart, 2010.

(27)   R. E. Goozner, M. M. Labes. Lytropic liquid crystals of D-(2)-octylammonium chloride. *Molecular Crystals and Liquid Crystals* **1985**, *116*, 309–317.

(28)   K. Radley, G. J. Lilly. A 2D-NMR Investigation of Deuterated Chiral Dopants in Amphiphilic Cholesteric Liquid Crystals. *Langmuir* **1997**, *13*, 3575–3578.

(29)   K. Radley, H. Cattey. An inversion of chirality at a chiral micelle surface. *Molecular Crystals and Liquid Crystals* **1994**, *250*, 167–175.

(30)   M. Pape, K. Hiltrop. Influence of lyotropic nematic host phases on the twisting power of chiral dopants. *Molecular Crystals and Liquid Crystals* **1997**, *307*, 155–173.

(31)   A. Agar, A. Gök. Comparison of the pitch induced from lamellar and nematic regions in some lyotropic liquid crystal systems. *Liquid Crystals* **1998**, *24*, 369–373.

(32)   R. S. Pindak, C. C. Huang, J. T. Ho. Divergence of cholesteric pitch near a smectic A transition. *Physical Review Letters* **1974**, *32*, 43–46.

(33)   S. H. Chen, J. J. Wu. Divergence of cholesteric pitch near the smectic-A transition in some cholesteryl nonanoate binary mixtures. *Molecular Crystals and Liquid Crystals* **1982**, *87*, 197–209.

(34)   A. Dequidt, P. Oswald. Zigzag instability of a chi disclination line in a cholesteric liquid crystal. *European Physical Journal E* **2006**, *19*, 489–500.

(35)   M. A. Osipov. Theory for cholesteric ordering in lyotropic liquid crystals. *Nuovo Cimento della Societa Italiana di Fiscina* **1988**, *10*, 1249–1262.

(36)   S. Gläßel. *Röntgenkleinwinkelstreuung an induzierten cholesterischen Phasen lyotroper Flüssigkristalle*. Diplomarbeit: Stuttgart, 2009.

(37)   J. Partyka, K. Hiltrop. On chirality induction in lyotropic nematic liquid crystals. *Liquid Crystals* **1996**, *20*, 611–618.

(38)   H. Baessler, M. M. Labes. Helical twisting power of steroidal solutes in cholesteric mesophases. *Journal of Chemical Physics* **1970**, *52*, 631–637.

(39)   H. Stegemeyer. *Berichte der Bunsen-Gesellschaft* **1974**, *78*, 860.

(40)   H. Finkelmann, H. Stegemeyer. Description of cholesteric mixtures by an extended Goossens theory. *Berichte der Bunsen-Gesellschaft* **1974**, *78*, 869–874.

(41)   J. Boos. *Kinetische Untersuchungen zur chiralen Induktion cholesterischer Phasen in lyotropen Flüssigkristallen*. Diplomarbeit: Stuttgart, 2009.

(42)   H. G. Kuball, T. Hofer. *Chirality in liquid crystals*; Springer: New York, 2001.

(43)   H. Stegemeyer, K. J. Mainusch. Induzierung von optischer Aktivität und Zirkulardichroismus in nematischen Phasen durch chirale Moleküle. *Naturwissenschaften* **1971**, *58*, 599–602.

(44)   L. Onsager. The effects of shapes on the interaction of colloidal particles. *Ann. N. Y. Acad. Sci.* **1949**, *51*, 627–659.

(45)   C. F. Dietrich, P. Rudquist, K. Lorenz, F. Gießelmann. Chiral Structures from Achiral Micellar Lyotropic Liquid Crystals under Capillary Confinement. *Langmuir* **2017**, *33*, 5852–5862.

(46)   D. L. Nelson, M. M. Cox. *Lehninger principles of biochemistry*; W.H. Freeman; Macmillan Higher Education: New York NY, Houndmills, Basingstoke, 2017.

(47)   H. Baumgärtel, E. U. Franck, W. Grünbein, H. Stegemeyer. *Topics in Physical Chemistry: Volume 3: Liquid Crystals*; Steinkopff-Verlag: Darmstadt, 1994.

(48)   A. S. Vasilevskaya, E.V. Generalova, S.S. Anatolii. Chromonic Mesophases. *Russ. Chem. Rev.* **1989**, *58*, 904–916.

(49)   P. J. Collings, A. J. Dickinson, E. C. Smith. Molecular aggregation and chromonic liquid crystals. *Liquid Crystals* **2010**, *37*, 701–710.

(50)   S. W. Tam-Chang, L. Huang. Chromonic liquid crystals: properties and applications as functional materials. *Chem. Commun.* **2008**, *17*, 1957–1967.

(51)   H.-S. Park, K. Whin-Woong, L. Tortora, Y. Nastishin, D. Finotello, S. Kumar, O. D. Lavrentovich. Self-Assembly of Lyotropic Chromonic Liquid Crystal Sunset Yellow and Effects of Ionic Additives. *J. Phys. Chem. B* **2008**, *112*, 16307–16319.

(52)   P. J. Collings, J. N. Goldstein, E. J. Hamilton, B. R. Mercado, K. J. Nieser, M. H. Rega. The nature of the assembly process in chromonic liquid crystals. *Liq. Cryst. Rev.* **2015**, *3*, 1–27.

(53)   H. S. Park, S. W. Kang, L. Tortora, S. Kumar, O. D. Lavrentovich. *Langmuir* **2011**, *27*, 4164–4175.

(54)   J. Lydon. Chromonic liquid crystalline phases. *Liquid Crystals* **2011**, *38*, 1663–1681.

(55)   C. G. Baumann, S. B. Smith, V. A. Bloomfield, C. Bustamante. Ionic effects on the elasticity of single DNA molecules. *Proc. Natl. Acad. Sci. U.S.A.* **1997**, *94*, 6185–6190.

(56)   Shuang Zhou. *Lyotropic Chromonic Liquid Crystals: From Viscoelastic Properties to Living Liquid Crystals*. Dissertation: Kent, USA, 2016.

(57)   S. Pieraccini, S. Masiero, A. Ferrarini, G. P. Spada. Chirality transfer across lenght-scales in nematic liquid crystals: fundamentals and applications. *Chem. Soc. Rev.* **2011**, *40*, 258–271.

(58)   C. Tschierske. Mirror symmetry breaking in liquids and liquid crystals. *Liquid Crystals* **2018**, *45*, 2221–2252.

(59)   M. Urbanski, C. G. Reyes, J. Noh, A. Sharma, Y. Geng, V. Subba Rao Jampani, J. P. F. Lagerwall. Liquid crystals in micron-scale droplets, shells and fibers. *Journal of physics. Condensed matter an Institute of Physics journal* **2017**, *29*, 133003.

(60)   A. Nych, U. Ognysta, J. Muševič, D. Seč, M. Ravnik, S. Žumer. Chiral bipolar colloids from nonchiral chormonic liquid crystals. *Phys Rev E* **2014**, *89*, 62502.

(61)   J. Jeong, L. Kang, Z. S. Davidson, P. J. Collings, T. C. Lubensky, A. G. Yodh. Chiral Structures from Achiral Liquid Crystals in Cylindrical Capillaries. *Proc. Natl. Acad. Sci. U.S.A.* **2015**, *112*, E1837-E1844.

(62) J. Jeong, Z. S. Davidson, P. J. Collings, T. C. Lubensky, A. J. Yodh. Chiral symmetry breaking and surface faceting in chromonic liquid crystal droplets with giant elastic anisotropy **2014**, *111*, 1742–1747.

(63) K. Nayani, J. Fu, R. Chang, J. O. Park, M. Srinivasarao. Using chiral tactoids as optical probes to study the aggregation behavior of chromonics. *Proc. Natl. Acad. Sci. U.S.A.* **2017**, *114*, 3826–3831.

(64) K. Nayani, R. Chang, J. Fu, P. W. Ellis, A. Fernandes-Nieves, J. O. Park, M. Srinivasarao. Spontaneous Emergence of Chirality in Achiral Lyotropic Chromonic Liquid Crystals Confined to Cylinders. *Nature Commun.* **2015**, *6*, 8067.

(65) L. Tortora, O. D. Lavrentovich. Chiral symmetry breaking by spatial confinement in tactoidal droplets of lyotropic chromonic liquid crystals. *Proc. Natl. Acad. Sci. U.S.A.* **2011**, *108*, 5163–5168.

(66) Z. S. Davidson, L. Kang, J. Jeong, T. Still, P. J. Collings, T. C. Lubensky, A. G. Yodh. Chiral structures and defects of lyotropic chromonic liquid crystals induced by saddle-splay elasticity. *Phys Rev E* **2015**, *91*, 050501/1-050501/5.

(67) S. Zhou, Y. A. Nastishin, M. M. Omelchenko, L. Tortora, V. G. Nazarenko, O. P. Boiko, T. Ostapenko, T. Hu, C. C. Almasan, S. Sprunt, J. T. Gleeson, O. D. Lavrentovich. Elasticity of lyotropic chromonic liquid crystals probed by director reorientation in a magnetic field. *Physical Review Letters* **2012**, *109*, 37801.

(68) S. Zhou, K. Neupane, Y. A. Nastishin, A. R. Baldwin, S. V. Shiyanovskii, O. D. Lavrentovich, S. Sprunt. Elasticity, viscosity, and orientational fluctuations of a lyotropic chromonic nematic liquid crystal disodium cromoglycate. *Soft matter* **2014**, *10*, 6571–6581.

(69) R. B. Meyer. On the existence of even indexed disclinations in nematic liquid crystals. *The Philosophical Magazine: A Journal of Theoretical Experimental and Applied Physics* **1973**, *27*, 405–424.

(70) C. Williams, Y. Bouligand. Fils et Disinclinaisons dans un Nématique en Tube capillaire. *J. Phys. (Paris)* **1974**, *35*, 589–593.

(71)   A. Saupe. Disclinations and Properties of the Directorfield in Nematic and Cholesteric
        Liquid Crystals. *Molecular Crystals and Liquid Crystals* **1973**, *21*, 211–238.

(72)   S. Kralj, S. Žumer. Saddle-Splay Elasticity of Nematic Structures Confined to a
        Cylindrical Capillary. *Phys. Rev. E: Stat. Phys., Plasmas, Fluids, Relat. Interdiscip.
        Top.* **1995**, *51*, 366–379.

(73)   P. E. Cladis, M. Kléman. Non-singular disclinations of strength S = +1 in nematics. *J.
        Phys. (Paris)* **1972**, *33*, 591–598.

(74)   C. E. Williams, P. Pieranski, P. E. Cladis. Nonsingular S = +1 Screw Disclination Lines
        in Nematics. *Physical Review Letters* **1972**, *29*, 90–92.

(75)   A. Shams, X, Y., J. O. Park, M. Srinivasarao, A. D. Rey. Theoretical predictions of
        disclination loop growth for nematic liquid crystals under capillary confinement.
        Physical review. E, Statistical, nonlinear, and soft matter physics **2014**, *90*, 42501.

(76)   G. P. Crawford, D. W. Allender, J. W. Doane. Surface elastic and molecular-anchoring
        properties of nematic liquid crystals confined to cylindrical cavities. *Phys. Rev. A* **1992**,
        45, 8693–8708.

(77)   W. H. de Jeu. *Physical properties of liquid crystalline materials*; Gordon and Breach
        Science Publishers, Inc.: New York, 1980.

(78)   *Light and Color - Optical Birefringence | Olympus Life Science.* https://www.olympus-
        lifescience.com/en/microscope-resource/primer/lightandcolor/birefringence/. Accessed
        29 May 2019.

(79)   *Michel-Lévy_interference_colour_chart_(21257606712).*
        https://upload.wikimedia.org/wikipedia/commons/6/64/Michel-
        L%C3%A9vy_interference_colour_chart_%2821257606712%29.jpg.   Accessed   28
        May 2019.

(80)   C. Görgens. *Strukturelle Charakterisierung lyotroper Mesophase, insbesondere
        lyotrop-nematischer   und   lyotrop-cholesterischer   Phasen,   mittels
        Röntgenkleinwinkeluntersuchungen in binären, ternären und quaternären Systemen.*
        Dissertation: Dresden, 1996.

(81)   M. C. Mauguin. Sur les cristaux liquides de Lehmann. *Bull. Soc. Fr. Mineral. Cristallogr.* **1911**, *34*, 71–117.

(82)   M. Kleman, O. D. Lavrentovich. *Soft Matter Physics: An Introduction*; Springer New York: New York, NY, 2004.

(83)   T. Turiv, I. Lazo, A. Brodin, B. I. Lev, V. Reiffenrath, V. G. Nazarenko, O. D. Lavrentovich. Effect of collective molecular reorientations on Brownian motion of colloids in nematic liquid crystal. *Science (New York, N.Y.)* **2013**, *342*, 1351–1354.

(84)   A. Shams, X. Yao, J. O. Park, M. Srinivasarao, A. D. Rey. Theory and modeling of nematic disclination branching under capillary confinement. *Soft matter* **2012**, *8*, 11135–11143.

(85)   G. S. Ranganath. Twist Disclinations in Elastically Anisotropic Nematic Liquid Crystals. *Molecular Crystals and Liquid Crystals* **1982**, *87*, 187–195.

(86)   G. S. Ranganath. Defects in Liquid Crystals. *Curr. Sci.* **1990**, *59*, 1106–1124.

(87)   O. D. Lavrentovich. Nematic Liquid Crystals: Defects. In *Encyclopedia of Materials: Science and Technology*, 2nd ed.; K. H. J. Buschow, R. W. Cahn, M. C. Flemings, B. Ilschner, E. J. Kramer, S. Mahajan, Ed.; Elsevier Science Ltd.: New York, 2001, pp. 6071–6076.

(88)   G. S. Ranganath. Energetics of Disclinations in Liquid Crystals. *Molecular Crystals and Liquid Crystals* **1983**, *97*, 77–94.

(89)   S. L. Anisimov, I. E. Dzyaloshinskii. A new type of disclination in liquid crystals and the stability of disclinations of various types. *Soviet. Phys. JETP* **1973**, *36*, 774–779.

(90)   D. W. Allender, G. P. Crawford, J. W. Doane. Determination of the Liquid-Crystal Surface Elastic Constant. *Physical Review Letters* **1991**, *67*, 1442–1445.

(91)   Stefanie Gläßel. *Röntgenkleinwinkelstreuung an induzierten cholesterischen Phasen lyotroper Flüssigkristalle.* Diplomarbeit: Stuttgart, 2009.

(92) M. Pape. *Über die Solubilisierung chiraler Gastmoleküle und den Mechanismus der chiralen Induktion in lyotropen flüssigkristallinen Gast-Wirt-Systemen.* Dissertation: Paderborn, 2000.

(93) E. Figgemeier, K. Hiltrop. Quantified chirality, molecular similarity, and helical twisting power in lyotropic chiral nematic guest/host systems. *Liquid Crystals* **1999**, *26*, 1301–1305.

(94) H.-S. Kitzerow, B. Liu, F. Xu, P. P. Crooker. Effect of chirality on liquid crystals in capillary tubes with parallel and perpendicular anchoring. *Phys Rev E* **1996**, *54*, 568–575.

(95) S. T. Lagerwall. *Ferroelectric and Antiferroelectric Liquid Crystals*; Wiley-Vch: Weinheim, 1999.

(96) Y. Ouchi, H. Takano, H. Takazoe, A. Fukuda. Zig-Zag Defects and Disclinations in the Surface-Stabilized Ferroelectric Liquid Crystals. *Jap. J. appl. Phys.* **1988**, *27*, 1–7.

(97) N. A Clark, T. P. Rieker. Smectic-C "chevron", a planar liquid-crystal defect: Implications for the surface-stabilized ferroelectric liquid-crystal geometry. *Phys Rev A* **1988**, *37*, 1053–1056.

(98) J. S. Patel, J. W. Goodby. Alignment of liquid crystals which exhibit cholesteric to smectic-C* phase transitions **1986**, *59*, 2355–2360.

(99) I. Dierking, F. Gießelmann, J. Schacht, P. Zugenmaier. Horizontal chevron configurations in ferroelectric liquid crystal cells induced by high electric fields. *Liquid Crystals* **1995**, *19*, 179–187.

(100) F. Gießelmann, P. Zugenmaier. Coupled Director and Layer Reorientation in Layer Tilted Ferroelectric Smectic Liquid Crystal Cells. *Molecular Crystals and Liquid Crystals Science and Technology. Section A. Molecular Crystals and Liquid Crystals* **1993**, *237*, 121–143.

(101) T. P. Rieker, N. A. Clark, D. S. Parmer, E. B. Sirota, C. R. Safinya. "Chevron" local layer structure in surface-stabilized ferroelectric smectic-C cells. *Physical Review Letters*, *59*, 2658–2661.

(102) P. W. Atkins. *Physikalische Chemie*; Wiley-Vch: Weinheim, 1988.

(103) D. Melzer, F. R. N. Nabarro. Cols and noeuls in a nematic liquid crystal with homeotropic cylindrical boundary. *Philos. Mag.* **1977**, *35*, 907–915.

(104) C. Carlini, F. Ciardelli, P. Pino. *Makromol. Chem.* **1968**, *119*, 244–248.

(105) M. M. Green, M. P. Reidy, R. d. Johnson, G. Darling, D. J. O'Leary, G. Willson. Macromolecular stereochemistry: the out-of-proportion influence of optically active comonomers on the conformational characteristics of polyisocyanates. The sergeants and soldiers experiment. *J. Am. Chem. Soc.* **1989**, *111*, 6452–6454.

(106) S. J. George, Z. Tomović, M. M. J. Smulders, T. F. A. de Greef, P. E. L. G. Leclère, E. W. Meijer, A. P. H. J. Schenning. Helicity induction and amplification in an oligo(p-phenylenevinylene) assembly through hydrogen-bonded chiral acids. *Angewandte Chemie (International ed. in English)* **2007**, *46*, 8206–8211.

(107) A. R. A. Palmans, J. A. J. M. Vekemans, E. E. Havinga, E. W. Meijer. Sergeants-and-Soldiers Principle in Chiral Columnar Stacks of Disc-Shaped Molecules withC3 Symmetry. *Angew. Chem. Int. Ed. Engl.* **1997**, *36*, 2648–2651.

(108) A. J. Wilson, M. Masuda, R. P. Sijbesma, E. W. Meijer. Chiral Amplification in the Transcription of Supramolecular Helicity into a Polymer Backbone. *Angew. Chem.* **2005**, *117*, 2315–2319.

(109) R. Stannarius. Elastic Properties of Nematic Liquid Crystals. In *Handbook of Liquid Crystals*, 2nd ed.; J. W. Goodby, P. J. Collings, T. Kato, C. Tschierske, H. Gleeson, P. Raynes (Hrsg.), Eds.; Wiley-Vch: Weinheim, 2014, pp. 60–90.

(110) V. Fréedericksz, A. Repiewa. Theoretisches und Experimentelles zur Frage nach der Natur der anisotropen Flssigkeiten. *Z. Physik* **1927**, *42*, 532–546.

(111) V. Fréedericksz, V. Zolina. Double refraction of thin layers of anisotropic liquids in a magnetic field, and the force orienting these layers. *Z. Kristallogr.* **1931**, *79*, 255–267.

(112) V. Fréedericksz, V. Zwetkoff. Über die Einwirkung des Elektrischen Feldes auf Anisotrope Flüssigkeiten, Orientierung der Flüssigkeit im Elektrischen Felde. *Acta Physiochim URSS* **1935**, *3*, 895.

(113) H. F. Gleeson. Light Scattering from Liquid Crystals. In *Handbook of Liquid Crystals*, 2nd ed.; J. W. Goodby, P. J. Collings, T. Kato, C. Tschierske, H. Gleeson, P. Raynes (Hrsg.), Eds.; Wiley-Vch: Weinheim, 2014, pp. 699–718.

(114) E. Miraldi, L. Trossi, P. T. Valabrega, C. Oldano. Absolute Measurements of the Elastic Constants of Nematic Liquid Crystals by Light Scattering. *Il nuovo cimento* **1981**, *66B*.

(115) Y. A. Nastishin, K. Neupane, A. R. Baldwin, O. D. Lavrentovich, S. Sprunt. Elasticity and Viscosity of a Lyotropic Chromonic Nematic Studied with Dynamic Light Scattering. *electronic-Liquid Crystal Communications* **2008**.

(116) V. G. Taratuta, A. J. Hurd, R. B. Meyer. Light-Scattering Study of a Polymer Nematic Liquid Crystal. *Physical Review Letters* **1985**, *55*, 246–249.

(117) L. Lucchetti, T. P. Fraccia, F. Ciciulla, T. Bellini. Non-linear optical measurement of the twist elastic constant in thermotropic and DNA lyotropic chiral nematics. *Scientific reports* **2017**, *7*, 4959.

(118) J. Bajc, G. Hillig, A. Saupe. Determination of elastic constants and rotational viscosity of micellar liquid crystals by conductivity measurements. *J. Chem. Phys.* **1997**, *106*, 7372–7377.

(119) M. Cui, J. R. Kelly. Temperature Dependence of Visco-Elastic Properties of 5CB. *Molecular Crystals and Liquid Crystals* **1999**, *331*, 49–57.

(120) S. D. Lee, R. B. Meyer. Light scattering measurements of anisotropic viscoelastic coefficients of a main-chain polymer nematic liquid crystal. *Liquid Crystals* **1990**, *7*, 15–29.

(121) M. Pregelj. *Seminar: Light Scattering on Liquid Crystals*: University of Ljubljana, Faculty of Mathematics and Physics, Department of Physics, 2005.

(122) Daniel Krüerke. *Experimentelle Untersuchungen nematischer und cholesterischer Phasen diskotischer Fssigkristalle.* Dissertation: Berlin, 1999.

(123) H. Dilger. *Skript zum Physikalisch-Chemischen Praktikum im Master:* Stuttgart, 2013.

(124) W. Schärtl. *Light Scattering from Polymer Solutions and Nanoparticle Dispersions*; Springer: Berlin/Heidelberg, 2007.

(125) C. B. McKitterick, N. L. Erb-Satullo, N: D. LaRacuente, A. J. Dickinson, P. J. Collings. Aggregation Properties of the Chromonic Liquid Crystal Benzopurpurin 4B. *J. Phys. Chem. B* **2010**, *114*, 1888–1896.

(126) K. Hata, Y. Takanishi, I. Nishiyama, J. Yamamoto. Sofenting of twist elasticity in the swollen smectic C liquid crystal. *EPL* **2917**, *120*, 56001.

(127) M. B. L. Santos, M. A. Amato. Orientational diffusivities measured by Rayleigh scattering in a lyotropic calamitic nematic liquid crystal phase: the backflow problem revisited. *European Physical Journal B* **1999**, *7*, 393–400.

(128) M. B. L. Santos, E. A. Oliveira, A. M. F. Neto. Rayleigh sacttering of a new lyotropic nematic liquid crystal system: crossover of propagative and diffusive behaviour. *Liquid Crystals* **2000**, *27*, 1485–1495.

(129) C. L. Risi, A. M. F. Neto. Dynamic light scattering and viscosity measurements in a ternary and quaternary discotic lyotropic nematic liquid crystal: Tuning the backflow with salt. *Phys Rev E* **2013**, *88*, 22506.

(130) *Ericksen-Leslie theory for nematic liquid crystals.* https://www.newton.ac.uk/files/seminar/20130723150016301-153704.pdf. Accessed 2 June 2019.

(131) F. M. Leslie. Continuum theory for nematic liquid crystals. *Continuum Mech. Thermodyn* **1992**, *4*, 167–175.

(132) D. R. Lide (Hrsg.). *CRC Handbook of Chemistry and Physics: Physical Constants of Organic Compounds,* 90th ed.; CRC Press/Taylor and Francis: Boca Raton, FL, 2010.

(133) P. Rudquist, C. F. Dietrich, A. G. Mark, F. Giesselmann. Chirality Detection Using Nematic Liquid Crystal Droplets on Anisotropic Surfaces. *Langmuir the ACS journal of surfaces and colloids* **2016**, *32*, 6140–6147.

(134) J. Martin, R. Cano. Light-propagation in cholesteric liquid-crystals in a domain including inversion range. *Nouv. Rev. Opt.* **1976**, *7*, 265–273.

(135) T. Lubensky, A. Harris, R. Kamien, G. Yan. Chirality in liquid crystals: From microscopic origins to macroscopic structure. *Ferroelectrics* **1998**, *212*, 1–20.

(136) H. G. Kuball, H. Brning. Helical twisting power and circular dichroism as chirality observations: The intramolecular and intermolecular chirality transfer. *Chirality* **1997**, *9*, 407–423.

(137) H. G. Kuball. From Chiral Molecules to Chiral Phases 1 Comments on the Chirality of Liquid Crystalline Phases. *Liquid Crystals Today* **1999**, *9*, 1–7.

(138) A. Harris, R. Kamien, T. Lubensky. Molecular chirality and chiral parameters. *Rev. Mod. Phys.* **1999**, *71*, 1745–1757.

(139) G. Solladié, R. Zimmermann. Liquid Crystals: A Tool for Studies on Chirality. *Angewandte Chemie (International ed. in English)* **1984**, *23*, 348–362.

(140) S. A. Issaenko, A. B. Harris, T. Lubensky. Quantum theory of chiral interactions in cholesteric liquid crystals. *Phys Rev E* **1999**, *60*, 578.

(141) R. Beradi, H. G. Kuball, R. Memmer, C. Zannoni. Chiral induction in nematics: A computer simulation study. *J. Chem. Soc., Faraday Trans.* **1998**, *94*, 1229–1234.

(142) J. Straley. Ordered phases of a liquid of biaxial particles. *Phys Rev A* **1974**, *10*, 1881–1887.

(143) A. Ferrarini, A. Gottarelli, P. L. Nordio, G. Spada. Determination of absolute configuration of helicenes and related biaryls from calculation of helical twisting powers by the surface chirality model. *J. Chem. Soc., Perkin Trans. 2* **1999**, *2*, 411–417.

(144) H. Kamberaj, M. Osipov, R. Low, M. Neal. Helical twisting power and chirality indices. *Mol. Phys.* **2004**, *102*, 431–446.

(145) M. Solymosi, R. Low, M. Grayson, M. Neal. A generalized scaling of a chiral index for molecules. *J. Chem. Phys.* **2002**, *116*, 9875–9881.

(146) S. Becke, S. Haller, M. Osipov, F. Giesselmann. Correlation between the molecular chirality index and the spontaneous polarisation in series of smectic C* liquid crystals. *Mol. Phys.* **2010**, *108*, 573–582.

(147) H.-S. Kitzerow, C. Bahr. *Chirality in Liquid Crystals*; Springer-Verlag New York Inc: New York, NY, 2001.

(148) B. L. Feringa. The Art of Building Small: From Molecular Switches to Molecular Motors. *J. Org. Chem.* **2007**, *72*, 6635–6652.

(149) D. Walba, L. Eshdat, E. Korblova, R. Shao, N. Clark. A general method for measurement of enantiomeric excess by using electrooptics in ferroelectric liquid crystals. *Angew. Chem.* **2007**, *119*, 1495–1497.

(150) N. Kapernaum, D. Walba, E. Korblova, C. Zhu, C. Jones, Y. Shen, N. Clark, F. Giesselmann. On the Origin of the 'Giant' Electroclinic Effect in a 'De Vries'-Type Ferroelectric Liquid Crystal Material for Chirality Sensing Applications. *Chem Phys Chem* **2009**, *10*, 890–892.

(151) M. Osipov, H. G. Kuball. Helical twisting power and circular dichroism in nematic liquid crystals doped with chiral molecules. *European Physical Journal E* **2001**, *5*, 589–598.

(152) G. Gray, D. McDonnell. Relationship between helical twist sense, absolute-configuration and molecular-structure for non-sterol cholesteric liquid-crystals. *Molecular Crystals and Liquid Crystals* **1976**, *34*, 211–217.

(153) H. Stegemeyer, R. Meister, H. J. Altenbach, D. Szewczyk. Ferroelectricity of induced S*C phases with novel chiral dopants. *Liquid Crystals* **1993**, *14*, 1007–1019.

(154) M. Osipov, H. Stegemeyer, A. Sprick. Molecular origin of ferroelectricity in induced smectic- C* liquid crystalline phases. *Phys. Rev. E* **1996**, *54*, 6387–6403.

(155) Nematic and Chiral Nematic Liquid Crystals. In *Handbook of Liquid Crystals*, 2nd ed.; J. W. Goodby, P. J. Collings, T. Kato, C. Tschierske, H. Gleeson, P. Raynes (Hrsg.), Eds.; Wiley-Vch: Weinheim, 2014; Vols. 3.

(156) M. Grandjean. Sur l'existence de plans différenciés équidistants normaux à l'axe optique dans les liquides anisotropes (cristaux liquides). *C. R. Acad. Sci.* **1921**, *172*, 71–74.

(157) R. Cano. Interprétation des discontinuités de Grandjean. *Bull. Soc. Fr. Mineral. Cristallogr.* **1968**, *91*, 20–27.

(158) S. Pieraccini, A. Ferrarini, G. P. Spada. Chiral doping of nematic phases and its application to the determination of absolute configuration. *Chirality* **2008**, *20*, 749–759.

(159) S. Suh, K. Joseph, G. Cohen, J. Patel, S. Lee. Precise determination of the cholesteric pitch of a chiral liquid crystal in a circularly aligned configuration. *App. Phys. Lett.* **1997**, *70*, 2547–2549.

(160) Y. Y. Tzeng, S. W. Ke, C. L. Ting, A. Y. G. Fuh, T. H. Lin. Axially symmetric polarization converters based on photo-aligned liquid crystal films. *Opt. Express* **2008**, *16*, 3768–3775.

(161) J. Goodby, I. Saez, S. Cowling, V. Görtz, M. Draper, A. Hall, S. Sia, G. Cosquer, S. Lee, E. Raynes. Transmission and amplification of information and properties in nanostructured liquid crystals. *Angewandte Chemie (International ed. in English)* **2008**, *47*, 2754–2787.

(162) E. Raynes. The use of bowed reverse twist disclination lines for the measurement of long pitch lengths in chiral nematic liquid crystals. *Liquid Crystals* **2006**, *33*, 1215–1218.

(163) R. Yamaguchi, S. Sato. Determination of nematic liquid crystal (NLC) orientation by observing NLC droplets on alignment surfaces. *Jpn. J. Appl. Phys.* **1996**, *35*, L117-L119.

(164) O. D. Lavrentovich. Topological defects in dispersed liquid crystals, or words and worlds around liquid crystal drops. *Liquid Crystals* **1998**, *24*, 117–125.

(165) G. Volovik, O. D. Lavrentovich. Topological dynamics of defect: boojums in nematic drops. *J. Exp. Theor. Phys.* **1983**, *58*, 1159–1166.

(166) H. Stegemeyer, H. J. Kersting, W. Kuczynski. Helical Twisting Power of Induced Twisted Smectic C Phases. *Berichte der Bunsen-Gesellschaft Phys. Chem.* **1987**, *91*, 3–7.

(167) G. Barbero, R. Barberi. Critical thickness of a hybrid aligned nematic liquid crystal cell. *J. Phys. (Paris)* **1983**, *44*, 609–616.

(168) L. Gvozdovskyy, I. Terenetskaya, V. Reshetnyak. Dissolution of steroid crystals in a nematic droplet. *Proc. SPIE* **2012**, *5257*, 102–109.

(169) N. Schnabel. *Photoisomerization and chirality effects in doped nematic liquid crystals under confined geometries.* Bachelor thesis: Stuttgart, 2019.